Springer Proceedings in Materials

Volume 22

Springer Proceedings in Materials publishes the latest research in Materials Science and Engineering presented at high standard academic conferences and scientific meetings. It provides a platform for researchers, professionals and students to present their scientific findings and stay up-to-date with the development in Materials Science and Engineering. The scope is multidisciplinary and ranges from fundamental to applied research, including, but not limited to:

- Structural Materials
- Metallic Materials
- Magnetic, Optical and Electronic Materials
- Ceramics, Glass, Composites, Natural Materials
- Biomaterials
- Nanotechnology
- Characterization and Evaluation of Materials
- Energy Materials
- Materials Processing

To submit a proposal or request further information, please contact one of our Springer Publishing Editors according to your affiliation:

European countries: **Mayra Castro** (mayra.castro@springer.com)
India, South Asia and Middle East: **Priya Vyas** (priya.vyas@springer.com)
South Korea: **Smith Chae** (smith.chae@springer.com)
Southeast Asia, Australia and New Zealand: **Ramesh Nath Premnath** (ramesh.premnath@springer.com)
The Americas: **Michael Luby** (michael.luby@springer.com)
China and all the other countries or regions: **Mengchu Huang** (mengchu.huang@springer.com)

This book series is indexed in **SCOPUS** database.

N. Madhusudhana Rao · Giribabu Lingamallu ·
Mangilal Agarwal
Editors

Advanced Nanomaterials and Their Applications

Select Proceedings of ICANA 2022

Springer

Editors
N. Madhusudhana Rao
School of Advanced Sciences
VIT-AP University
Amaravati, Andhra Pradesh, India

Mangilal Agarwal
Department of Mechanical and Energy Engineering
Indiana University-Purdue University Indianapolis
Indianapolis, IN, USA

Giribabu Lingamallu
Department of Polymers and Functional Materials
CSIR-Indian Institute of Chemical Technology
Hyderabad, India

ISSN 2662-3161 ISSN 2662-317X (electronic)
Springer Proceedings in Materials
ISBN 978-981-99-1615-3 ISBN 978-981-99-1616-0 (eBook)
https://doi.org/10.1007/978-981-99-1616-0

This Springer imprint is published by the registered company Springer Nature Singapore Pte Ltd.
The registered company address is: 152 Beach Road, #21-01/04 Gateway East, Singapore 189721, Singapore

Contents

Ion Beam Synthesis of Germanium Nanocrystals—A Fluence Dependence Study 1
V. Saikiran, G. Neelima, N. Srinivasa Rao, and A. P. Pathak

Graphene Oxide–Agar–Agar Hydrogel for Efficient Removal of Methyl Orange from Water 9
Tufan Ghosh

Numerical Analysis of Novel $Cs_2AuBiCl_6$-Based Double Perovskite Solar Cells with Graphene Oxide as HTL—A SCAPS-1D Simulation ... 21
Titu Thomas, Davis Johny, and B. Sudakshina

Nanoscopic Pd(II)-Based Complexes with Poly-Ether Functionalized Ligand: The Crown Ether Analog 33
Debakanta Tripathy, Soumya Lipsa Rath, Srabani Srotwosini Mishra, and Dillip Kumar Chand

Preparation of Hydrotalcite–$CdPS_3$ Hybrid Solid from the Exfoliated Inorganic Nanosheets 49
Rajesh Chalasani

Deposition Time-Dependant Properties of PbS Thin Films 57
Srinivasa Reddy Tippasani, S. Vijaya Krishna, and M. C. Santhosh Kumar

Investigation on Surface Trap Characteristics of Water-Diffused Al–Epoxy Nanocomposites 67
Chillu Naresh and Ramanujam Sarathi

A Study on Impact of Hydrophobic Effect on Al_2O_3 Coated Glass by Sol–Gel Dip Coating Method for Automobile Windshield Application 77
K. Chandru and R. Elansezhian

Design and Analysis of Chalcogenide GeAsSe Waveguide for Dispersion Properties 87
V. Hitaishi, K. Jayakrishnan, and Nandam Ashok

Detection of Pathological Conditions in Nail Samples Using Laser-Induced Breakdown Spectroscopy 97
K. Rithika, R. Sowmya, G. Rithick kumar, M. Thangaraja, Pauline John, and V. Sathiesh Kumar

A Review of mRNA Vaccines with the Aid of Lipid Nanoparticles 111
Simran Saikia, Shreya Barman, S. Sudhimon, M. Mukesh Kumar, G. Shanmugasundaram, and J. Sudagar

Metal–Organic Framework for Antibiotic Sensing Application 125
Krupa U. Patel, Dashrathbhai B. Kanzariya, and Tapan K. Pal

Metal Organic Framework (MOF)-Based Membranes for Separation Applications 135
Dashrathbhai B. Kanzariya, Krupa U. Patel, Rudra Desai, and Tapan K. Pal

Control of Dissolved Oxygen in Wastewater Treatment Plant Using NN Adaptive PID Controller 147
S. M. Tharani, A. Ganesh Ram, and M. Vijayakarthick

About the Editors

Dr. N. Madhusudhana Rao is a professor at the Department of Physics, School of Advanced Sciences, VIT-AP University, Andhra Pradesh, India. His current research areas include dilute magnetic semiconductors (II–VI, IV–VI), transparent conducting oxides, and luminescent materials for optoelectronics. He obtained a Ph.D. in physics from Sri Venkateswara University, India. Dr. Rao has completed three projects funded by various prestigious Indian agencies like DRDO and UGC-DAE-CSR. He has published over 75 research papers, 24 conferences/book/book chapters, and 2 Indian patents. Dr. Rao has guided five Ph.D. students who have completed their degrees. Dr. Rao organized the prestigious 59th DAE Solid State Physics Symposium in collaboration with BARC and BRNS, Department of Atomic Energy, Government of India, from 16 to 20 December in VIT University, Vellore.

Dr. Giribabu Lingamallu is a senior principal scientist at CSIR-Indian Institute of Chemical Technology, Hyderabad, India. He has more than 25 years of research experience in the area of materials chemistry that includes Excitonic Solar Cells (Dye-Sensitized Solar Cells/Perovskite Solar Cells), Non-linear Optics, and Donor–Acceptor systems for bio-mimicking natural photosynthesis. Dr. Lingamallu obtained his Ph.D. degree from the University of Hyderabad, India. He was a post-doctoral fellow at Central Queens Land University, Rockhampton, Australia, working in the area of photoinduced reactions in Porphyrin-Alicyclic compounds. From February 2001 to July 2003, he worked as a post-doctoral fellow at the University of Houston, USA, in the area of bi-metallic (di-Rhodium and di-Ruthenium) complexes synthesis, crystal analysis, and electrochemical studies. He moved to CSIR-Indian Institute of Chemical Technology, Hyderabad, India, in September 2003 as ad-hoc scientist, scientist, senior scientist, principal scientist, and now as senior principal scientist where his group was involved in establishing third-generation photovoltaic facilities that includes the development of low-cost efficiency materials (sensitizers, redox couples, semi-conducting oxide materials, electrode materials, hole transporting materials, etc.) and also device fabrication. In addition to this, his group is also working in the area of Donor–Acceptor systems for bio-mimicking natural photosynthesis, sensitizers for photodynamic therapy, and materials for non-linear optical studies.

Dr. Mangilal Agarwal is a full-time professor in the Department of Mechanical and Energy Engineering, Indiana University-Purdue University of Indianapolis (IUPUI), USA. He has an array of interdisciplinary research projects at IUPUI involving big data, data analytics, machine learning, analytical chemistry, and mechanical engineering. Specifically, his research consists of canine-inspired chemometric analysis of volatile organic compound (VOC) biomarkers, developing integrated nanosensor arrays that detect VOC biomarkers sensitively/selectively, and enhancing carbon-reinforced materials through electrospinning epoxy/carbon nanotube composite nanofibers. Dr. Agarwal's research has been well received and funded through numerous grants from the National Science Foundation (NSF), National Institute of Health (NIH), Department of Defense (DoD), and the industry. He has previously served as the associate director for Research and Development for the IUPUI Office of the Vice Chancellors for Research and is currently the director of Integrated Nanosystems Development Institute (INDI) at IUPUI, which strives to advance nanotechnology research, education, and outreach. From innovation to translation, research within INDI's two focus areas of Bionanotechnology and Nanoenergy takes not only the advantage of unique campus strengths but also the expertise of the surrounding industry.

Ion Beam Synthesis of Germanium Nanocrystals—A Fluence Dependence Study

V. Saikiran, G. Neelima, N. Srinivasa Rao, and A. P. Pathak

Abstract The synthesis of Ge nanocrystals (NCs) by using ion implantation method is reported here along with the results from different spectroscopic and microscopic characterizations such as Rutherford backscattering spectroscopy (RBS), X-ray diffraction (XRD), Raman spectroscopy, photoluminescence (PL), and atomic force microscopy (AFM). Various fluences of 1 MeV Ge ions have been implanted into SiO_2, and then, as-implanted samples were annealed using rapid thermal annealing system for the synthesis of Ge NCs. The Ge NCs presence was confirmed from Raman spectroscopy and XRD measurements. The low-fluence implanted sample did not show any signature of Ge NCs, whereas Ge NCs presence has been observed in the high-fluence implanted sample after annealing. The mechanism of Ge NCs formation in the as-implanted samples after annealing has been discussed.

Keywords Ion beam synthesis · Ion implantation · Rapid thermal annealing · Ge nanocrystals · Raman · Photoluminescence

1 Introduction

Nanocrystals (NCs) embedded in a dielectric matrix are of huge interest due to their wide variety of applications as photonic and electronic devices. Among metal and semiconductor NCs, semiconductor NCs are found to have potential applications in non-volatile memories and optical devices [1–5]. Careful optimization of the physical properties of these NCs is vital for implementing them in device applications. Particularly, NCs of silicon (Si) and germanium (Ge) embedded in SiO_2 matrix have

V. Saikiran (✉) · G. Neelima
Department of Physics, School of Sciences, GITAM Deemed to be University, Visakhapatnam, Andhra Pradesh 530045, India
e-mail: svadaval@gitam.edu

N. S. Rao
Department of Physics, Malaviya National Institute of Technology Jaipur, Jaipur 302017, India

A. P. Pathak
School of Physics, University of Hyderabad, Hyderabad 500046, India

N. M. Rao et al. (eds.), *Advanced Nanomaterials and Their Applications*,
Springer Proceedings in Materials 22, https://doi.org/10.1007/978-981-99-1616-0_1

received an enormous amount of attention because of their applications in integrated nanoelectronics, nanoflash memory, and nanodevices [6, 7]. Ge NCs due to their relatively deep quantum well have a broad range of applications, such as a charge storage node in memory devices, photodetectors, and light emitters [8–10]. Ge shows strong quantum confinement compared to Si because of low effective mass and high dielectric permittivity of its charge carriers [11]. But the real hassle in NCs properties and their applications is the control in the size and shape of NCs and their distribution. Ion beams have a special place to rectify this issue, and as reported, ion beams can be used for both the synthesis of NCs and also for the modification of pre-existing NCs [12–14]. There were various methods reported for Ge NCs synthesis in dielectric matrix like sputtering (DC, RF, and atom beam), ion implantation, evaporation (thermal and electron beam), and laser-based synthesis [15–18]. Among them, ion implantation is a more suitable technique for the site-selective synthesis of NCs by implanting the desired ions into dielectric matrix, and it has been widely used for NCs fabrication. The ion fluence during implantation will have a crucial effect on the formation of NCs embedded in any dielectric matrix. Generally, after the implantation of the desired ions, either with rapid thermal annealing (RTA) treatment or with swift heavy-ion (SHI) irradiation, the embedded NCs can be synthesized. The additional advantages of this method over other methods are that the implantation can be done at isolated locations and selective sites. In the present work, Ge ions were implanted in SiO_2 with various fluences having 1 MeV energy. The as-implanted samples were annealed using RTA at 800 °C. The formation of Ge NCs was characterized by using XRD and Raman spectroscopy methods. The changes in surface topography after the implantation were studied by atomic force microscopy (AFM). The photoluminescence (PL) spectroscopy confirms the optical emission from the Ge NCs and the oxygen-related defects created during implantation.

2 Experimental Details

Ge ion implantation into thermally grown SiO_2 films having energy of 1 MeV was performed using Tandem accelerator facilities of IGCAR, Kalpakkam. The fluences were varied from 1×10^{16} to 1×10^{17} ions/cm^2 with implantation performed under room temperature conditions. The as-implanted films were annealed using RTA process for about 5 min at a fixed temperature of 800 °C under N_2 atmosphere. The as-implanted and RTA annealed films have been characterized with XRD and Raman spectroscopy (with 514.5 nm excitation wavelength) for checking the formation of Ge NCs in SiO_2. RBS has been used for quantifying the concentration of implanted Ge in the SiO_2 films. PL spectra at room temperature were recorded using WITec Alpha 300 spectrometer with 355 nm excitation wavelength. The topography of the films has been studied using AFM model SPA 400 of SPI 3800 probe station of Seiko Instruments.

3 Results and Discussion

Figure 1 shows the RBS spectrum of the 1×10^{17} ions/cm^2 fluence Ge implanted in thermally grown SiO_2. It indicates that the implanted films consist of various elements Si, oxygen (O), and Ge in the sample. From the data, it is estimated that the concentration of Ge is to be 4 at% in the case of 1×10^{17} ions/cm^2 fluence sample. The other low-fluence implanted films show lesser Ge than these samples.

The XRD pattern of the RTA-treated Ge implanted SiO_2 films is shown in Fig. 2. We have not observed any crystalline Ge peaks from the XRD pattern of the as-implanted sample and also for the RTA-treated 1×10^{16} ions/cm^2 fluence sample. The samples with further higher fluence implantation after RTA annealing show (101), (113) crystalline peaks of Ge (JCPDS 03-065-0333), which is an indication (for the creation) of growth of Ge NCs as a result of implantation followed by RTA treatment [19–21].

Figure 3 represents the Raman spectra of the RTA-treated SiO_2 films implanted with 1 MeV Ge ions of different fluences. For comparison, the Raman data of the as-implanted sample of the fluence 1×10^{17} ions/cm^2 is also shown in the figure. The sample with this highest implanted fluence has a broad Raman peak around wavenumber 270 cm^{-1} which is related to the optical phonons of Ge–Ge vibrations [22–24]. The broadness in the peak indicates the amorphous nature of the as-implanted Ge ions into SiO_2 film at the room temperature. After annealing the RTA-treated samples at 800 °C, the Ge–Ge optical phonon-related peak becomes sharp around the wavenumber 301 cm^{-1} with an asymmetry toward the lower wavenumber

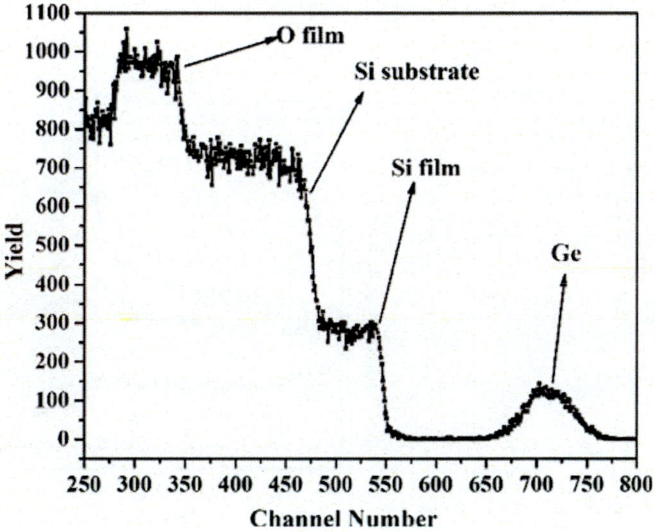

Fig. 1 RBS spectrum of the 1×10^{17} ions/cm^2 fluence Ge implanted into SiO_2 film

Fig. 2 XRD pattern of the RTA annealed films implanted with 1 MeV Ge ions of varied fluences

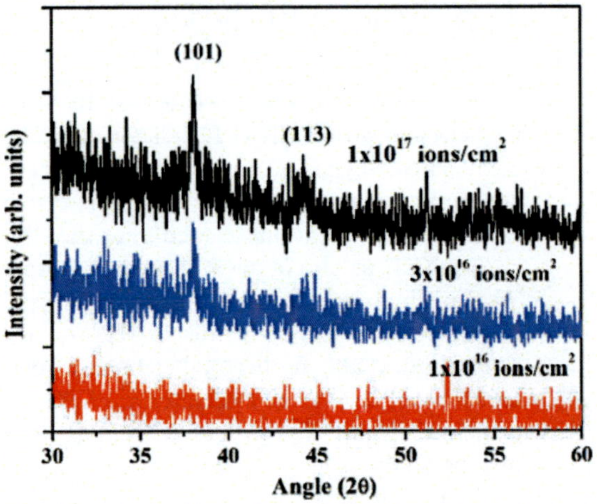

[24]. This sudden change after annealing indicates that the crystallization occurs and confirms the creation of Ge NCs in SiO_2 layers. The similar kind of sharp peak is observed in the case of other 3×10^{16} ions/cm^2 fluence-annealed sample also. As the implanted Ge was less in fluence compared to 1×10^{17} ions/cm^2 sample, this sample has broad and less intense Raman peak, whereas the 1×10^{16} ions/cm^2 fluence annealed sample did not show any signature of the Ge optical phonon peak before and after annealing. This may be due to the low atomic concentration of Ge at that fluence. The shift in the peak position from 301 cm^{-1} can be used to estimate the average size of the Ge NCs.

Fig. 3 Raman spectra of the as implanted (1×10^{17} ions/cm^2 fluence) sample in comparison with that of the 800 °C RTA-treated samples of different fluences of 1 MeV Ge ions

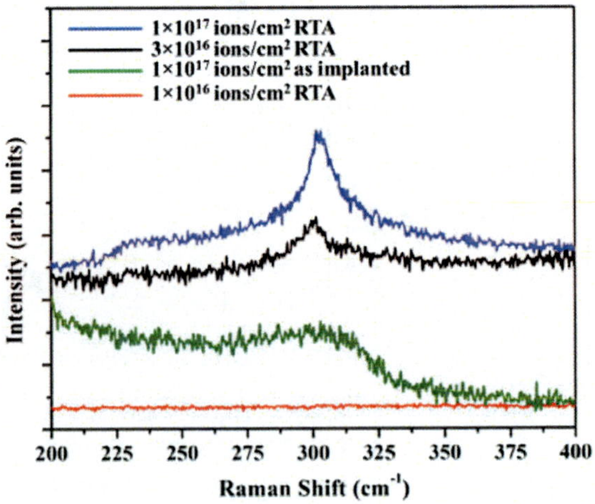

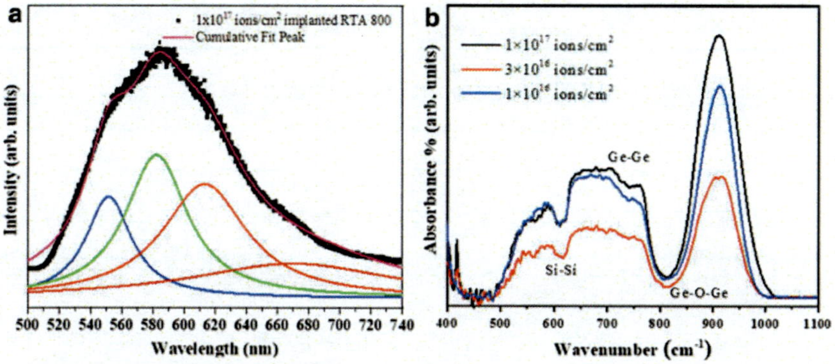

Fig. 4 **a** PL emission spectra of the RTA-treated sample implanted with 1×10^{17} ions/cm^2 fluence 1 MeV Ge ions along with the Lorentzian deconvoluted peaks and **b** FTIR spectra of the RTA-treated samples implanted with various fluences of 1 MeV Ge ions

Figure 4a shows the PL emission spectra of the RTA-treated sample of 1 MeV Ge ions implanted in SiO_2 films. The respective Lorentzian deconvoluted peaks are also presented in the figure. The sample was excited with a wavelength of 355 nm, and the spectra were recorded at room temperature. The PL emission spectra depict an emission in the range from 550 to 750 nm. The spectra have peak centers around 551, 582, 612, and 670 nm. The emission is spread in the green to red emission. These are due to the implantation induced oxygen related defects created in SiO_2 [25]. The peaks at 551 and 580 nm (green emission) may be due to the Ge NCs formed as a result of RTA treatment of the as-implanted samples [26].

The FTIR spectra of the RTA-treated sample of 1 MeV Ge ions implanted in SiO_2 films are shown in Fig. 4b. The FTIR spectra are used to understand the chemical bonding nature in the implanted samples. The sample indicates peaks related to Ge–Ge, Ge–O–Ge bonds present in the Ge implanted SiO_2 samples at wavenumbers around 610, 800 cm^{-1}, etc. [27, 28]. The FTIR spectra indicate the clear formation of Ge bonding and its presence in the SiO_2 films. The morphology of the implanted films was recorded using AFM, and the topography is presented in Fig. 5. The top surface of the SiO_2 film looks similar in the surface roughness before and after implantation of Ge ions, and also the surface appears smooth with an average roughness of 2–3 nm. These images indicate that the surface of SiO_2 did not get damaged as a result of implantation. However, the inner region gets modified, and defects are created inside SiO_2 which is evident from the PL emission results. The AFM images also clearly demonstrate that the segregation of the particles on the surface is more prominent in the 1×10^{17} ions/cm^2 fluence sample than the 3×10^{16} ions/cm^2 fluence sample.

The mechanism involved in the ion beam synthesis of Ge NCs as a result of Ge ion implantation and followed by RTA treatment can be explained as below. The implantation creates few islands of Ge atoms at a depth of around 500–600 nm in SiO_2 films as estimated from stopping and range of ions in matter (SRIM) simulations. Under the RTA treatment, these Ge atoms diffuse and agglomerate to form Ge NCs.

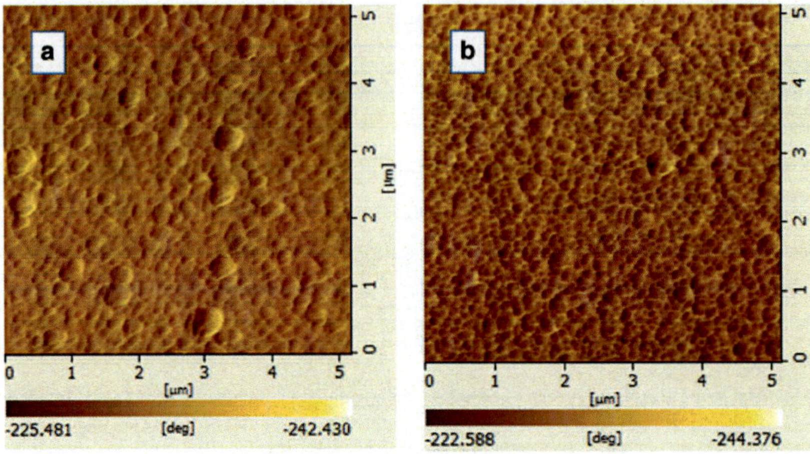

Fig. 5 AFM surface morphologies of SiO_2 films implanted with 1 MeV Ge ions of different fluences **a** 3×10^{16} and **b** 1×10^{17} ions/cm^2

During the annealing process, thermal energy is supplied to the SiO_2 films which are implanted with Ge ions. The growth of Ge grains through the SiO_2 results in the nucleation of Ge NCs in films. The average size and growth of the NCs strongly depend on the implanted ion fluence. The typical size and thereby physical properties of Ge NCs can be tuned by selecting desired energy and ion fluence during the ion beam synthesis method.

4 Conclusions

Ge NCs have been synthesized by the ion beam synthesis by using 1 MeV Ge implantation into SiO_2 and followed by RTA treatment. XRD and Raman measurements indicate the presence of Ge NCs. It is observed that the NCs are formed in high-fluence implanted sample and in the low-fluence samples they are not observed, thereby indicating that the implantation fluence plays a vital role on the synthesis of Ge NCs in this method. It is observed from PL emission experiments that the PL emission can be tuned based on the choice of the implantation fluence. The mechanism of Ge NCs formation in the as-implanted samples after annealing has been discussed.

Acknowledgements VS acknowledges funding received from UGC and SERB through grants UGC-F.30-456/2018(BSR) and SERB-SRG/2019/001830. APP acknowledges National Academy of Sciences, Prayagraj for award of NASI Senior Scientist Platinum Jubilee Fellowship. We are thankful to UGC NRC Centre of School of Physics at UOHYD for extending Raman and XRD facilities.

References

1. Choi WK, Chim WK, Heng CL, Teo LW, Ho V, Ng V, Antoniadis DA, Fitzgerald EA (2002) Appl Phys Lett 80:2014
2. Lim KY, Kim MC, Hong SH, Choi S, Kim KJ (2010) J Appl Phys 108:033708
3. Das K, Goswami MN, Mahapatra R, Kar GS, Dhar A, Acharya HN, Maikap S, Lee J-H, Ray SK (2004) Appl Phys Lett 84:1386
4. Leong WL, Lee PS, Mhaisalkar SG (2007) Appl Phys Lett 90:042906
5. Ray SK, Maikap S, Banerjee W, Das S (2013) J Phys D: Appl Phys 46:153001
6. Cosentino S, Mirabella S, Liu P, Le ST, Miritello M, Lee S, Crupi I, Nicotra G, Spinella C, Paine D, Terrasi A, Zaslavsky A, Pacifici D (2013) Thin Solid Films 548:551–555
7. Hong SH, Kim MC, Jeong PS, Choi S, Kim Y, Kim KJ (2008) Appl Phys Lett 92:093124
8. Park CJ, Cho KH, Yang WC, Cho HY, Choi SH, Elliman RG et al (2006) Appl Phys Lett 88:071916
9. Colace L, Masini G, Assanto G, Luan HC, Wada K, Kimerling LC (2000) Appl Phys Lett 76:1231
10. Carolan D (2017) Prog Mater Sci 90:128–158
11. Maeda Y (1995) Phys Rev B 5:1658
12. Baranwal V, Gerlach JW, Lotnyk A, Rauschenbach B, Karl H, Ojha S, Avasthi DK, Kanjilal D, Pandey AC (2015) J Appl Phys 118:134303
13. Araujo LL, Giulian R, Sprouster DJ, Schnohr CS, Llewellyn DJ, Johannessen B, Byrne AP, Ridgway MC (2012) Phys Rev B 85:235417
14. Saikiran V, Srinivasa Rao N, Devaraju G, Chang GS, Pathak AP (2013) Nucl Inst Meth Phys Res Sect B 315:161–164
15. Volodin VA, Rui Z, Krivyakin GK, Antonenko AK, Stoffel M, Rinnert H, Vergnat M (2018) Semiconductors 52:1178–1187
16. Lehninger D, Seidel P, Geyer M, Schneider F, Klemm V, Rafaja D, von Borany J, Heitmann J (2015) Appl Phys Lett 106:023116
17. Vadavalli S, Valligatla S, Neelamraju B, Dar MH, Chiasera A, Ferrari M, Desai NR (2014) Front Phys 2:57
18. Srinivasa Rao N, Pathak AP, Sathish N, Devaraju G, Saikiran V, Kulriya PK, Agarwal DC, Sai Saravanan G, Avasthi DK (2010) Solid State Commun 150:2122–2126
19. Kalimuthu V., Kumar P, Kumar M, Rath S (2018) Appl Phys A 124:712
20. Samavati A, Othaman Z, Ghosh SK, Dousti MR (2014) J Lumin 154:51–57
21. Rao NS, Dhamodaran S, Pathak AP, Kulriya PK, Mishra YK, Singh F, Kabiraj D, Pivin JC, Avasthi DK (2007) Nucl Inst Meth Phys Res B 264:249–253
22. Choi WK, Ng V, Ng SP, Thio HH, Shen ZX, Li WS (1999) J Appl Phys 86:1398
23. Campbell IH, Fauchet PM (1986) Solid State Commun 58:739–741
24. Guha S, Wall M, Chase LL (1999) Nucl Inst Meth Phys Res Sect B 147:367–372
25. Ray SK, Das K (2005) Opt Mater 27:948–952
26. Das K, Goswami MLN, Dhar A, Mathur BK, Ray SK (2007) Nanotechnology 18:175301
27. Ortiz MI, Rodriguez A, Sangrador J, Rodriguez T, Avella M, Jimenez J, Ballesteros C (2005) Nanotechnology 16:S197
28. Witanachchi S, Wolf PJ (1994) J Appl Phys 76:2185

Graphene Oxide–Agar–Agar Hydrogel for Efficient Removal of Methyl Orange from Water

Tufan Ghosh

Abstract We report the preparation of graphene oxide (GO)–agar–agar hydrogel and its utilization toward the removal of methyl orange (MO) dye from water. Graphene oxide–agar–agar composite hydrogel has been prepared from a mixture of chemically synthesized GO and agar–agar powder in hot water followed by cooling at room temperature. To study the dye removal property of the prepared composite hydrogel, we placed the GO–agar hydrogel into an aqueous solution of MO. The dye adsorption was apparent as the color of the solution started disappearing immediately after the addition of the GO–hydrogel. To quantify this effect further, the dye removal efficiency was evaluated and the results suggest that the removal efficiency could reach up to 99% depending on the pH of the medium and the dose of GO used. In an acidic medium (pH ~ 2), the removal efficiency was found to be relatively higher, compared to that of an alkaline medium. The removal efficiency was improved as the dose of GO in the hydrogel was increased. Further, the kinetics of the dye adsorption onto the GO surface has been examined by plotting the experimental data according to Lagergren pseudo-first-order and pseudo-second-order models. The results from these analyses suggest that the pseudo-second-order model gives a better fit for the experimental data, than the first-order model. Our results show that GO–agar gel can be efficiently used for the removal of various toxic dyes from wastewater.

Keywords Dye removal · Methyl orange · Graphene oxide · Wastewater treatment · Dye adsorption

1 Introduction

Water contamination due to the disposal of various unwanted materials is of global environmental concern to scientists [1, 2]. Due to the increased dumping of pollutants into water, the level of contamination has a constant rate of increase. The pollutants include organic dye, heavy metals, and oils [3–5]. Organic dyes have been widely

T. Ghosh (✉)
Department of Chemistry, School of Advanced Sciences, VIT-AP University, Amaravati, Andhra Pradesh 522237, India
e-mail: tufan.ghosh@vitap.ac.in

N. M. Rao et al. (eds.), *Advanced Nanomaterials and Their Applications*,
Springer Proceedings in Materials 22, https://doi.org/10.1007/978-981-99-1616-0_2

used in various industries to color various products. The dye contamination occurs when they are disposed of in water without appropriate discharge treatment. The dye contamination is of great concern since some of the degraded products from the used dyes show carcinogenic and toxic behavior. Heavy metals are of serious concern also since they have created environmental problems. Various methods have been developed for the removal of toxic dyes and metal ions from wastewater. Some examples of the removal methods are coagulation, liquid membrane separation, biological treatment and adsorption [6]. However, each method differs in its effectiveness of removal and cost of operation. Among these methods, the adsorption method is widely accepted in the water treatment industry due to its ability to complete the removal of various types of dyes and ease of the process [7, 8]. The material which works best for the dye removal should have a high specific surface area, chemically stable, environment friendly, and of course are low cost. While, the high specific area helps an absorbent to adsorb more amount of pollutant dyes per unit area, chemical stability is required to utilize the material in different chemical environments. Various carbon-based materials such as activated carbon, carbon nanotube, and recently graphene have been used for water purification [9–11].

Graphene has a high specific surface area (2630 $m^2\ g^{-1}$) along with other unique properties such as its large area planer structure, high pore volume, presence of oxygen containing groups, and high chemical and thermal stability. The adsorption of organic dye usually occurs via π–π interaction between the dye and graphene sheets [12]. Recent studies have shown that graphene and graphene-based materials are used in the removal of various dyes and metal ions [13, 14]. Various dyes including methylene blue, rhodamine B, acridine orange, etc. have been successively removed from water by using GO as an adsorbent [15–18]. Recently, graphene-based composite materials also have been introduced as an efficient absorbent for various pollutants. For example, reduced graphene oxide-based hydrogels have been prepared and shown that the composite gel can be utilized for the efficient removal of methylene blue and rhodamine B [19]. Yuan et al. have covalently modified GO with poly(amidoamine) and shown that this composite can remove heavy metal ions such as Pb^{2+}, Cu^{2+}, Zn^{2+}, and Cr^{3+} [20]. A few studies also report the removal of methyl orange from wastewater using graphene oxide composites [21–24]; nonetheless, none of these utilizes a low-cost and easy-to-prepare graphene–gel composite.

Herein we report the preparation of a graphene oxide–agar–agar-based hydrogel for the removal of methyl orange dye from water. Our systematic investigation demonstrates that the efficiency of the removal of MO is extremely higher when the medium is acidic, and the efficiency reduces at an alkaline medium. The removal efficiency reported here is ~99%, suggesting that this composite could be a very good candidate for MO removal from wastewater. Furthermore, the dose of GO also has a prominent effect on the removal efficiency which is demonstrated in the later section of this manuscript. Finally, the kinetic of the dye adsorption was monitored by plotting the experimental data using various models, and the results suggest that pseudo-second-order kinetic model could be used to best fit the experimental data.

2 Materials and Methods

2.1 Materials

Natural graphite powder (−325 mesh) was purchased from Alfa Aesar. Methyl orange was purchased from SD Fine Chemical Limited, India, and agar–agar bacto powder was purchased from Fischer Scientific, India. All materials were used as received without further purification. Double-distilled water was used as a solvent for carrying out the experiments.

2.2 Preparation of GO–Agar–Agar Composite Hydrogel

Graphene oxide was prepared according to a modified Hummers's method. The details of the synthesis procedure have been reported in our previous publications [25]. Agar–agar gel has been prepared by cooling down a hot solution of it. In a typical method, 0.3 g of agar–agar bacto powder was dissolved in 20 mL of double-distilled water at 95 °C. Then the solution was allowed to cool down to room temperature to form agar–agar gel. The gel was semi-transparent in nature. For the preparation of GO–agar composite hydrogel, GO powder was added to the hot aqueous solution of agar–agar and ultrasonicated for 10 min, followed by cooling at room temperature. The formation of the gel was confirmed by the tube inversion method (Fig. 2b). A black color gel was formed which was used for further dye removal experiments.

2.3 Performance of the Dye Adsorption Experiments

The dye adsorption/removal experiments were carried out using a UV-visible absorption spectrophotometer (JASCO V660) at room temperature (298 K). A quartz cuvette with a path length of 1 cm was used for the UV-visible absorption measurements. For the adsorption measurements, a stock solution of MO (6.20 mg/L) was prepared first and as a stock solution, and the remaining experiments were carried out by further diluting the stock solution. In a typical method, GO–agar composite gel was added to a 50 mL aqueous solution of MO (6.2 mg/L), and the absorption of the dye was monitored by measuring the absorbance of the solution at 500 nm (λ_{max} for MO solution) at different times in order to estimate the dye removal efficiency. The specific adsorption amount of MO at equilibrium (Q_e) and at time t (Q_t), and the efficiency of removal (E) have been determined according to the following well-known equations [15]:

$$Q_e = \frac{C_0 - C_e}{C_{\text{GO}}} \tag{1}$$

$$Q_t = \frac{C_0 - C_t}{C_{GO}} \tag{2}$$

$$\text{And } E = \left(1 - \frac{C_e}{C_0}\right) \times 100\% \tag{3}$$

where C_0 (mg/L) is the initial concentration of the MO, C_e (mg/L) is the equilibrium concentration of MO, and C_{GO} (g/L) is the concentration of GO.

The effect of pH of the medium on the removal efficiency of MO was studied by adjusting the pH of the solution in a pH range of 3–10. The pH of the solution was adjusted by adding either dilute HCl or NaOH solutions. The effect of GO dose on the removal efficiency of MO was investigated by stirring a 50 mL MO solution (6.2 mg/L) with different amounts of GO (0.0246–0.0991 g) separately.

3 Results and Discussion

3.1 UV–Visible Absorption Characteristic of MO

The MO molecule changes its structure depending on the pH of the medium. Figure 1a represents the anionic form and one of the protonated forms. The protonated form shown here could again be in equilibrium with other forms (not shown here) [26]. Some of these structural changes are also reflected in its electronic absorption spectrum recorded using a UV-visible absorption spectrophotometer. Figure 1b presents the UV-visible absorption spectra of an aqueous solution of MO at various pH. The MO molecules in an aqueous medium show a strong absorption peak around 465 nm, and the peak maximum remains unchanged even in strongly alkaline medium. However, at a significantly acidic medium (pH ~ 2.5), the absorption maximum shifts to 500 nm. These observations are in line with previous experiments that MO molecules being a charged dye change their structure in an aqueous solution. In the next sections, we will examine how these structural changes influence the dye removal efficiency in presence of GO–agar gel.

3.2 Removal of MO from Water Using GO–Agar Hydrogel

The adsorption experiments were carried out by adding GO–agar–agar gel to an aqueous solution of MO (6.2 mg/L). The adsorption of MO onto GO-based gel was monitored by measuring the absorption of MO at 500 nm using a UV-visible spectrophotometer. An aliquot was taken from the reaction vessel, and the absorption at different times was measured. It was found that the absorption of MO at 500 nm decreased gradually with time, indicating a reduction in the concentration of MO in

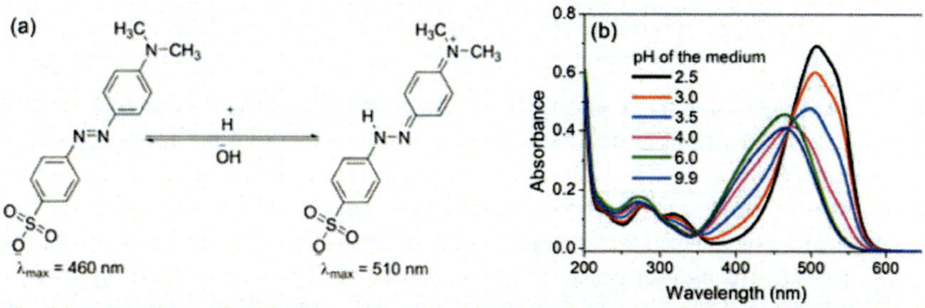

Fig. 1 **a** Structural transformations of methyl orange (MO) in acidic and alkaline medium; **b** UV-visible absorption spectra of MO solution at different pH: 2.5 (black), 3 (orange), 3.5 (blue), 4 (magenta), 6 (olive), and 9 (navy)

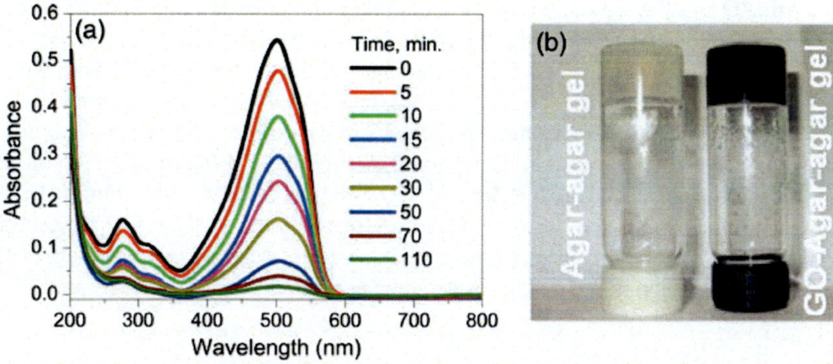

Fig. 2 **a** UV-visible absorption spectra of MO at different times after addition of the GO–agar–agar gel. **b** Photographs of semi-transparent agar–agar gel (left) and GO–agar gel (right)

the solution. This decrease in absorption of MO could mean either degradation or adsorption both of which can lead to a decrease in the number of MO molecules in the solution. Since the UV-visible absorption spectra maintain their original shape (to that of MO solution alone) even after the addition of GO–Agar gel, we assign the decrease in absorption to the adsorption of MO onto the surface of GO sheets. Figure 2a illustrates the UV-visible absorption spectra of MO taken at different times after the addition of GO–agar gel. From the absorbance values at 500 nm at different times after addition of the GO–agar gel, we have calculated the C_t and C_e which were further used for the calculation of the removal efficiency of MO according to Eq. (3).

3.3 *Effect of pH and GO Dose on Removal of MO*

As the MO dye is an anionic dye and GO also contains various oxygen-containing functional groups which are easily ionizable, the pH of the medium could play an important role in the adsorption process, i.e., removal efficiencies as well as the rate of adsorption. To understand the role of the pH of the medium on removal efficiencies, we have performed the adsorption experiments at three different pH values (pH: 3, 6, and 10). The plot for pH-dependent removal efficiency is shown in Fig. 3a. The results suggest that the efficiency of removal of MO reaches ~98% in an acidic medium, whereas the value of removal efficiency is much lower (~80%) in an alkaline medium. This is presumably because of the fact that MO and GO are structurally sensitive to the pH of the medium and undergo the protonation–de-protonation process. The adsorption of MO onto GO occurs mainly via π–π interaction between benzene rings of MO and the aromatic rings of GO. At the alkaline medium, MO exists in anionic form, at the same time carboxylic acid and hydroxyl groups are deprotonated in GO, and there will be a huge negative charge on the surface of GO. This will lead to repulsion between the π–electrons and, hence weaken the interaction.

Next, in order to evaluate the effect of GO dose on the removal efficiency, we have also carried out the adsorption experiments at different dosages of GO. The results suggest that as the dose of GO was increased the dye removal efficiency also increased as the dose of GO was increased up to 99% (at GO concentration of 1.982 g/L). The increase in efficiency is obvious because, as the amount of GO was increased the amount of the surface area as well as active sites increased.

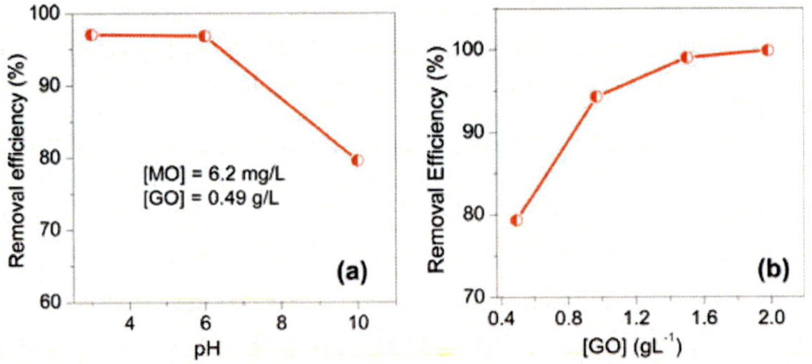

Fig. 3 **a** Plot of removal efficiency of MO versus pH of the medium. [MO] = 6.2 mg/L, [GO] = 0.49 g/L. **b** Plot of removal efficiency of MO versus [GO] in g/L, [MO] = 6.2 mg/L, [GO] = 0.492–1.982 g/L

3.4 Kinetics of Adsorption

Explaining the kinetics of adsorption is essential to understand the fundamentals of any adsorption process. Various models have been reported to understand the adsorption mechanism. Lagergren pseudo-first-order model, pseudo-second-order model, Elovich kinetic equation, parabolic diffusion model, etc. are examples of such models used in adsorption studies. Mostly, Lagergren pseudo-first-order and pseudo-second-order models have been used for analyzing the adsorption kinetics in recent the past. We have verified the adsorption kinetics data of MO by using Lagergren pseudo-first-order and pseudo-second-order models. The Lagergren pseudo-first-order model can be expressed as below [15]:

$$\ln(Q_e - Q_t) = \ln Q_e - K_1 t \tag{4}$$

and the pseudo-second-order model can be written as:

$$\frac{t}{Q_t} = \frac{1}{K_2 Q_e^2} + \frac{1}{Q_e} \tag{5}$$

where Q_e and Q_t are the maximum amounts of MO (mg/g) adsorbed on GO–agar–agar composite gel at equilibrium and at a given time t (min), respectively, and K_1 (min^{-1}) and K_2 ($\text{g mg}^{-1}\ \text{min}^{-1}$) are respective rate constants of adsorption process for the pseudo-first-order and pseudo-second-order models.

First, we have fitted the data according to the pseudo-first-order model. The plot of $\ln(Q_e - Q_t)$ versus time does not show an expected linear decrease (Fig. 4a). The value of the correlation coefficient (0.6532) suggests that the data fitting is not appropriate. Thus, we fitted the data with a pseudo-second-order model. The plot was drawn for t/Q_t versus time, which shows a good straight-line fitting (Fig. 4b). The correlation coefficient value for the fitting in this case was 0.9982 suggesting a good fitting. From the fitting, we have calculated the value of maximum specific adsorption (43.47 mg/g) which is closer to that of the experimentally obtained value (55.56 mg/g). The value of the maximum specific adsorption was compared with the values reported in the literature. This suggests that the obtained value for the present system is better than that of the previous report where a graphene sponge was utilized for the removal of MO from water [27]. The following table presents the adsorption kinetics parameters for the adsorption of MO by GO–agar–agar composite gel (Table 1). However, it is worth to note here that these kinetics models do not give any indication about the adsorption mechanisms.

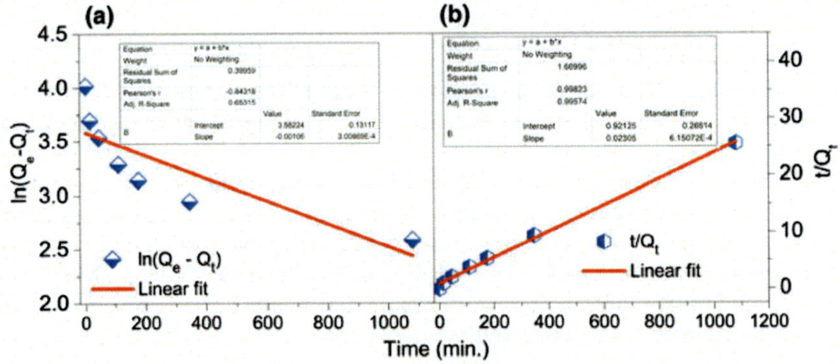

Fig. 4 Fitting of the experimental data with **a** pseudo-first-order and **b** pseudo-second-order kinetics model, for the adsorption of MO onto GO–agar–agar composite gel. [MO] = 24.5 mg/L, [GO] = 0.1 g/L

4 Conclusions

Graphene–agar–agar composite hydrogel has been prepared by mixing GO and agar–agar powder in hot water followed by cooling to room temperature. The GO–agar–agar gel shows excellent adsorption toward organic toxic dyes, such as methyl orange. A detailed spectroscopic study shows that pH of the medium and GO dosage have a significant role on the removal efficiency of MO. The removal efficiency was maximum in an acidic medium and decreased, as the medium was made alkaline. The maximum removal efficiency of MO has been calculated as 98% in an acidic medium. The kinetic analysis suggests that the adsorption mechanism follows a pseudo-second-order model. Also, we have determined the rate constant of MO adsorption using UV-visible absorption spectroscopy. The present study shows that graphene oxide-based gel material could be efficiently utilized for the removal of various toxic dyes from water.

Table 1 Adsorption kinetics parameter for MO adsorption on GO–agar–agar composite gel

Initial concentration, Co (mg/L)	Pseudo-first-order model			Pseudo-second-order model			$Q_{e,\text{exp}}$ (mg/g)
	K_1 (min^{-1})	$Q_{e,\text{cal}}$ (mg/g)	R^2	K_2 ($\text{g mg}^{-1}\ \text{min}^{-1}$)	$Q_{e,\text{cal}}$ (mg/g)	R^2	
24.5	0.00106	12.025	0.6532	0.000574	43.47	0.9982	55.56

References

1. Montgomery MA, Elimelech M (2007) Water and sanitation in developing countries: including health in the equation. Environ Sci Technol 41(1):17–24
2. Khan S, Malik A (2014) Environmental and health effects of textile industry wastewater. In: Malik A, Grohmann E, Akhtar R (eds) Environmental deterioration and human health: natural and anthropogenic determinants. Springer, Netherlands, pp 55–71
3. Robinson T, McMullan G, Marchant R, Nigam P (2001) Remediation of dyes in textile effluent: a critical review on current treatment technologies with a proposed alternative. Biores Technol 77(3):247–255
4. Mohammed AS, Kapri A, Goel R (2011) Heavy metal pollution: source, impact, and remedies. In: Khan MS, Zaidi A, Goel R, Musarrat J (eds) Biomanagement of metal-contaminated soils. Springer, Netherlands, pp 1–28
5. Fischer K, Fries E, Körner W, Schmalz C, Zwiener C (2012) New developments in the trace analysis of organic water pollutants. Appl Microbiol Biotechnol 94(1):11–28
6. Katheresan V, Kansedo J, Lau SY (2018) Efficiency of various recent wastewater dye removal methods: a review. J Environ Chem Eng 6(4):4676–4697
7. Ali I, Gupta VK (2006) Advances in water treatment by adsorption technology. Nat Protoc 1(6):2661–2667
8. Gupta VK, Suhas (2009) Application of low-cost adsorbents for dye removal—a review. J Environ Manag 90(8):2313–2342
9. Rai S, De A, Guin M, Singh NB (2022) Carbon materials for dye removal from wastewater. In: Muthu SS, Khadir A (eds) Textile wastewater treatment: sustainable bio-nano materials and macromolecules, vol 1. Springer Nature, pp 141–183
10. Moosavi S, Lai CW, Gan S, Zamiri G, Akbarzadeh Pivehzhani O, Johan MR (2020) Application of efficient magnetic particles and activated carbon for dye removal from wastewater. ACS Omega 5(33):20684–20697
11. Mohammadi N, Khani H, Gupta VK, Amereh E, Agarwal S (2011) Adsorption process of methyl orange dye onto mesoporous carbon material–kinetic and thermodynamic studies. J Colloid Interface Sci 362(2):457–462
12. Kyzas GZ, Deliyanni EA, Matis KA (2014) Graphene oxide and its application as an adsorbent for wastewater treatment. J Chem Technol Biotechnol 89(2):196–205
13. Liu X, Ma R, Wang X, Ma Y, Yang Y, Zhuang L, Zhang S, Jehan R, Chen J, Wang X (2019) Graphene oxide-based materials for efficient removal of heavy metal ions from aqueous solution: a review. Environ Pollut 252:62–73
14. Ramesha GK, Vijaya Kumara A, Muralidhara HB, Sampath S (2011) Graphene and graphene oxide as effective adsorbents toward anionic and cationic dyes. J Colloid Interface Sci 361(1):270–277
15. Liu T, Li Y, Du Q, Sun J, Jiao Y, Yang G, Wang Z, Xia Y, Zhang W, Wang K, Zhu H, Wu D (2012) Adsorption of methylene blue from aqueous solution by graphene. Colloids Surf, B 90:197–203
16. Yang S-T, Chen S, Chang Y, Cao A, Liu Y, Wang H (2011) Removal of methylene blue from aqueous solution by graphene oxide. J Colloid Interface Sci 359(1):24–29
17. El-Shafai NM, El-Khouly ME, El-Kemary M, Ramadan MS, Masoud MS (2018) Graphene oxide–metal oxide nanocomposites: fabrication, characterization and removal of cationic rhodamine B dye. RSC Adv 8(24):13323–13332
18. Sun L, Yu H, Fugetsu B (2012) Graphene oxide adsorption enhanced by in situ reduction with sodium hydrosulfite to remove acridine orange from aqueous solution. J Hazard Mater 203–204:101–110
19. Tiwari JN, Mahesh K, Le NH, Kemp KC, Timilsina R, Tiwari RN, Kim KS (2013) Reduced graphene oxide-based hydrogels for the efficient capture of dye pollutants from aqueous solutions. Carbon 56:173–182
20. Yuan Y, Zhang G, Li Y, Zhang G, Zhang F, Fan X (2013) Poly(amidoamine) modified graphene oxide as an efficient adsorbent for heavy metal ions. Polym Chem 4(6):2164–2167

21. Robati D, Mirza B, Rajabi M, Moradi O, Tyagi I, Agarwal S, Gupta VK (2016) Removal of hazardous dyes-BR 12 and methyl orange using graphene oxide as an adsorbent from aqueous phase. Chem Eng J 284:687–697
22. Yang Z, Ji S, Gao W, Zhang C, Ren L, Tjiu WW, Zhang Z, Pan J, Liu T (2013) Magnetic nanomaterial derived from graphene oxide/layered double hydroxide hybrid for efficient removal of methyl orange from aqueous solution. J Colloid Interface Sci 408:25–32
23. Filice S, D'Angelo D, Libertino S, Nicotera I, Kosma V, Privitera V, Scalese S (2015) Graphene oxide and titania hybrid Nafion membranes for efficient removal of methyl orange dye from water. Carbon 82:489–499
24. Wu S, Yu L, Xiao F, You X, Yang C, Cheng J (2017) Synthesis of aluminum-based MOF/graphite oxide composite and enhanced removal of methyl orange. J Alloy Compd 724:625–632
25. Ghosh T, Maity S, Prasad E (2014) Rate and mechanistic investigation of $Eu(OTf)_2$-mediated reduction of graphene oxide at room temperature. J Phys Chem B 118(20):5524–5531
26. He S, Sun X, Zhang H (2016) Influence of the protonation state on the binding mode of methyl orange with cucurbiturils. J Mol Struct 1107:182–188
27. Zhao J, Ren W, Cheng H-M (2012) Graphene sponge for efficient and repeatable adsorption and desorption of water contaminations. J Mater Chem 22(38):20197–20202

Numerical Analysis of Novel $Cs_2AuBiCl_6$-Based Double Perovskite Solar Cells with Graphene Oxide as HTL—A SCAPS-1D Simulation

Titu Thomas, Davis Johny, and B. Sudakshina

Abstract Perovskite solar cells (PSCs) are promising candidates to address today's energy crisis. The most challenging obstacles to the commercialization of PSCs are their volatility toward environmental conditions and the presence of lead (Pb). To solve this, the scientific community has come up with double perovskites having a general molecular formula of $A_2BB'X_6$. In this study, we introduce a novel structure $Cs_2AuBiCl_6$ as an absorber due to its non-toxic nature and stable performance. Also, the electron transport layer (ETL) and hole transporting layer (HTL) play key roles in the performance and stability of PSCs. But most of the widely used HTL and ETL materials are very costly and have complex synthesizing process. Here, we use a novel HTL material graphene oxide (GO) as HTL and ZnSe as ETL, both of them being cheap, non-toxic, and easily available. We investigate for the first time ever the performance of a ZnSe/$Cs_2AuBiCl_6$/GO structure using the Solar Cell Capacitance Simulator (SCAPS-1D) software. Results indicate that the optimized thickness was 1 μm for absorber and 0.1 μm for HTL. The device efficiency improved with increasing the shunt resistance to 50 Ω, while it deteriorated with series resistance. Finally, all the output parameters declined with the rise in the operating temperature beyond 300 K. The elicited results suggest that $Cs_2AuBiCl_6$ and GO can play a momentous role in achieving highly efficient lead-free, inorganic perovskite solar cells.

Keywords $Cs_2AuBiCl_6$ · Double perovskites · Graphene oxide · SCAPS · Numerical simulation

T. Thomas (✉)
Department of Physics, Nirmala College, Muvattupuzha, Kerala 686661, India
e-mail: tt@nirmalacollege.ac.in

D. Johny
Centre for Energy Materials, MG University, Priyadarshini Hills Kottayam 686560, India

B. Sudakshina
Department of Physics, T.M. Jacob Memorial Government College, Manimalakunnu 686662, India

N. M. Rao et al. (eds.), *Advanced Nanomaterials and Their Applications*,
Springer Proceedings in Materials 22, https://doi.org/10.1007/978-981-99-1616-0_3

1 Introduction

Perovskite solar cells are the main competitor for the widely used silicon solar cells. The efficiency has reached an astonishing 25% in only a short span of 10 years [1]. The optoelectronic properties of the perovskites like tunable direct band gap, long carrier diffusion length, high carrier lifetime, tolerance to various defects, etc. make them an ideal candidate for photovoltaic applications. But the perovskite was not devoid of disadvantages as well. Severe problems like rapid degradation on exposure to humidity and the presence of toxic components prevent the usage of perovskite solar cells at the commercial level [2].

Various attempts have been tried to ponder this problem like replacing the toxic lead with other elements like Sn but proved unsuccessful due to their lower solubility and crystallization properties [3] and also their tendency to convert to Sn^{4+} from Sn^{2+} states. Another solution was the heterovalent substitution of lead using Cs^{+} and Sb^{4+} or Bi^{3+} in the perovskite structure [4], but they also had low dimensionality leading to poor performance and stability issues [5]. Another approach is to replace lead with two cations, a monovalent metal cation and a trivalent metal cation, and such structures are called double perovskites or elpasolites and have been under study for about 30 years now [6, 7]. They can be represented as $AB_1B_2X_6$, where A is a cation, B_1 and B_2 are some anions, and X represent any halogen [8]. They have unique properties like high tolerance factor [9] high absorption, reduced carrier effective masses [10] band structure similar to that of lead halides [11] non-toxic and stiffness [12]. Due to all these favorable optoelectronic properties, double halides are now used not only for photovoltaic purposes but also in X-ray and UV detectors [13] laser applications [14] and photocatalysis [15], etc. More than 9000 double perovskite structures are possible theoretically though all of them are not stable [16, 17], and most of them have been widely studied but have not made any considerable improvement in performance [18]. In this article, we try a new double perovskite structure $Cs_2AuBiCl_6$. This unexplored double perovskite material has all the advantages of the double structure. And with a computed band gap of 1.12 eV [19], this material is highly efficient as per the Shockley–Queisser limit in the case of a single-junction solar cell and is comparable with other proven double perovskite-based photovoltaic devices [20, 21].

The hole transporting layer (HTL) and electron transport layer (ETL) also play a vital role in the performance of perovskite solar cells [22, 23]. Suitable HTL and ETL act as a path for the carriers to reach the respective electrodes. Most often hole transporting layers are generally organic polymers such as (2,2′,7,7′-Tetrakis[N,N-di(4-methoxyphenyl)amino]-9,9′-spirobifluorene] (Spiro-OMeTAD) [36], poly(triaryl amine) (PTAA) [37], poly(3-hexylthiophene-2,5-diyl) (P3HT) [38], and poly(3,4-ethylenedioxythiophene):poly(styrene sulfonate) (PEDOT:PSS) [24]. Most of these HTL materials are very expensive at the same time they are highly unstable. Recently, graphene and its derivatives like graphene oxide and reduced graphene oxide have started to revolutionize the optoelectronic industry with their unique properties [25,

26]. Here we introduce a novel hole transporting material graphene oxide (GO) which is non-toxic, cheap, widely available and has all suitable optoelectronic properties.

1.1 Simulation Methodology

Numerical simulation is a crucial part of any experiment saving a vast amount of time, and cost and giving a deep insight into the actual physical phenomenon playing inside. There are many simulation packages available in the market such as COMSOLE, MATLAB, SCAPS-1D, AMPS, etc. Here we use the SCAPS software developed at the Department of Electronics and Information System, University of Gent, Belgium [27]. The program uses three equations: the carrier continuity equations, the poisons equation, and the drift–diffusion Eqs. (1), (2), (3), (4) and (5) and solves these equations for various input parameters to give various device outputs [28].

$$\frac{\partial^2\varphi}{\partial x^2} = \frac{q}{\varepsilon}(n - p) \tag{1}$$

$$\frac{\partial n}{\partial \mathrm{t}} = (G - R) \tag{2}$$

$$\frac{\partial p}{\partial \mathrm{t}} = (G - R) \tag{3}$$

$$J_n = qD_n\frac{\partial n}{\partial \mathrm{x}} - q\mu_n n\frac{\partial \varphi}{\partial \mathrm{x}} \tag{4}$$

$$J_p = -qD_p\frac{\partial p}{\partial \mathrm{x}} - q\mu_p p\frac{\partial \varphi}{\partial \mathrm{x}} \tag{5}$$

where φ is the electric potential, q is the electronic charge, ε is the dielectric constant, n is the electron concentration, p is the hole concentration, J_n is the electron density, J_p is the hole current density, G is the carrier generation rate, and R is the carrier recombination rate. D_n and D_p are electron and hole diffusion coefficients, respectively, and μ_n and μ_p are the electron and hole mobility, respectively. The SCAPS is preferable to other simulation packages due to its user-friendly interface and flexibility in fixing various parameters. Materialistic properties like the thickness, band gap, electron affinity, carrier concentration, and other operating parameters like the temperature, applied voltage, frequency, and illumination can all be changed easily, and the results are available in a wide variety of formats. The device is represented as a stack of different layers in the software, and any device with up to 7 layers can be simulated using SCAPS-1D. Along with that, there are also provisions for front and back contacts and to determine the effect of various interfaces between these layers.

Table 1 Simulation parameters used in the numerical analysis

Properties	GO	$Cs_2AuBiCl_6$	ZnSe
Thickness (μm)	0.1	1	0.05
Bandgap (eV)	2.48	1.12	2.9
Electron affinity (eV)	2.3	4.5	4.09
Dielectric permittivity (relative)	10	3.56	10
Conduction band effective density of states (cm^{-3})	2.2E+18	2.17E+18	1.5E+18
Valence band effective density of states (cm^{-3})	1.8E+19	4.07E+18	1.8E+18
Electron mobility (cm^2/Vs)	26	10	50
Hole mobility (cm^2/Vs)	123	10	20
Donor density (cm^{-3})	0	1.0E+10	5.5E+18
Acceptor density (cm^{-3})	2.0E+18	1.0E+10	0
References	[29]	[30]	[31]

1.2 Device Structure and Simulation Parameters

In the present study, the proposed device has a structure $ZnSe/Cs_2AuBiCl_6/GO$. Table 1 gives the simulation parameters that we have used here. Most of them are taken from various published articles, and a few are by reasonable assumption. Band alignment is very important in the case of charge diffusion of heterojunction solar cells and determines the flow of charge carriers as shown in Fig. 1. The simulation is conducted under the illumination of AM1.5G with an intensity of 1000 mW/cm^2. In this work, the effect of the thickness of the absorber layer and the HTL layer was studied. Along with the material properties, operational conditions of the device such as the series and shunt resistance of the device and the effect of the operating temperature were also investigated.

2 Results and Discussions

2.1 Effect of Thickness of the Absorber Layer

The thickness of the perovskite solar cell is an important factor in terms of performance as well as in determining the cost of the device. The thickness of the $Cs_2AuBiBr_6$ layer was varied from 0.1 to 1 μm, and the results are elicited in Fig. 2. The efficiency of the device increased from 9 to 19%, and this can be explained as follows. If the thickness of the absorber layer is small, the depletion layer will be close to the back contact. And a portion of the incident photos may recombine at the back contact rather than at the depletion layer, thereby increasing the contact recombination probability. But with the increase in thickness of the absorber layer more

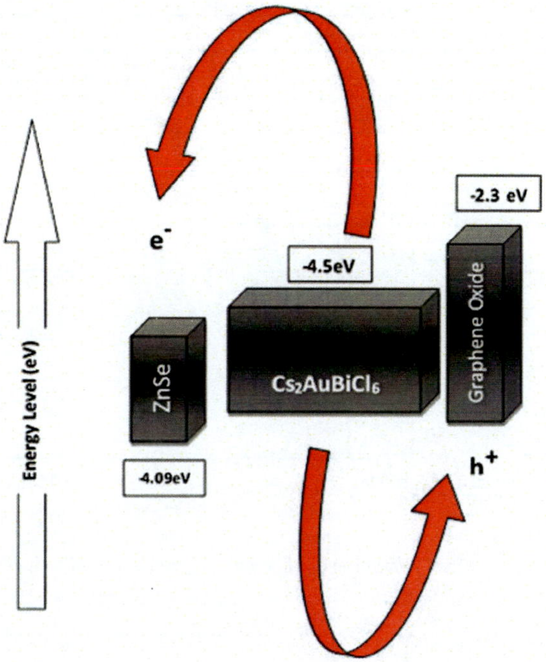

Fig. 1 Band diagram of the proposed structure

and more photons will reach deeply into the material and recombine at the depletion layer decreasing the recombination at the contacts and improving the device performance [32]. In the present study, the efficiency of the device increases with thickness, and then it remains saturated after a thickness of 0.5 μm. This occurs due to the phenomenon that with any further increase in the thickness of the absorber layer, the carriers fail to reach the depletion region and undergo recombination within the bulk itself.

2.2 Effect of Thickness of the HTL Layer

The variation in the HTL layer does not have much effect usually in perovskite solar cells. Investigation of the proposed structure also revealed similar results. The thickness was varied from 0.1 to 1 μm. The maximum efficiency obtained was 19%, and it fairly remained constant throughout the variation. The variation is shown in Fig. 3.

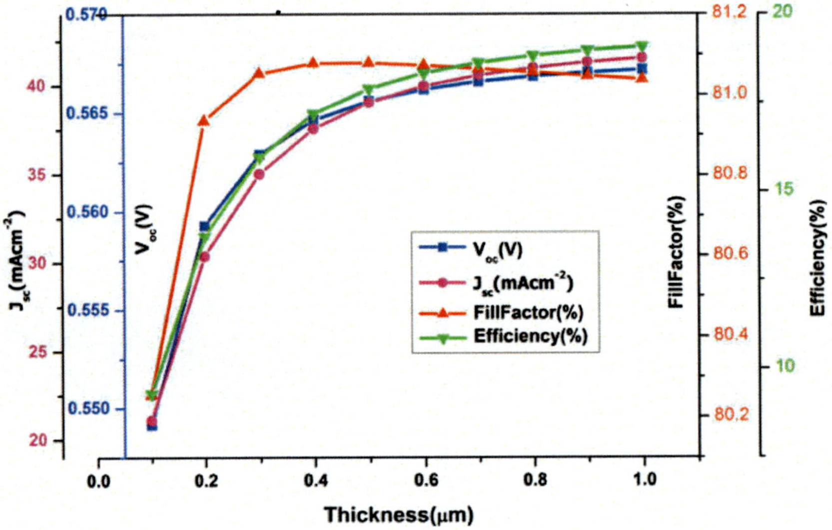

Fig. 2 Variation of photovoltaic parameters with change in absorber thickness

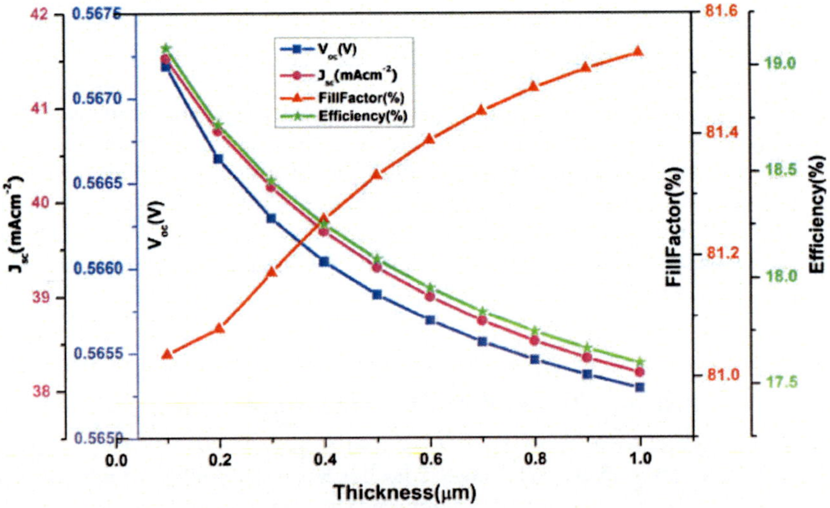

Fig. 3 Variation of photovoltaic parameters with change in HTL thickness

2.3 *Effect of Temperature*

Solar cells are outdoor devices and are continuously exposed to sunlight. So the effect of operating temperature was severely studied. We have changed the operating

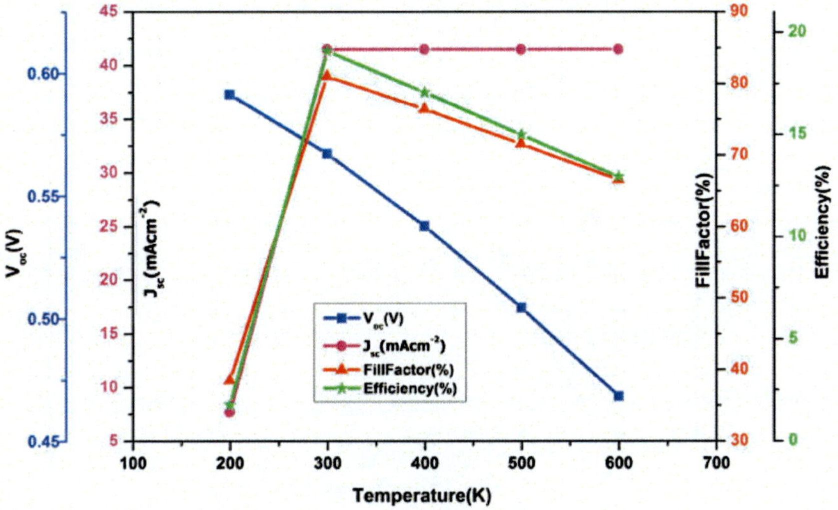

Fig. 4 Variation of photovoltaic parameters with change in temperature

temperature from 200 to 600 K. As shown in Fig. 4, the temperature has a huge impact on the various photovoltaic parameters. All the parameters declined rapidly with the change in temperature. The efficiency of the device dropped from 19% to almost 2% with a slight change in temperature drop from 300 to 200 K. Also, on further increasing the temperature beyond 300 K, the parameters drop and hence 300 K was fixed as the optimum temperature for the working of the proposed device.

2.4 Effect of Parasitic Resistance

The series and shunt resistances are the two resistance losses in a photovoltaic device that cannot be avoided and are generally called parasitic resistances. Among them, the series resistance is generally caused by the electrical connections between various layers and the contacts. The various electrical dissipations occurring at the ETL, HTL, and absorber layers increase the series resistances. Any other form of electrical resistance occurring in the device due to any design issues or any impurities causing carrier recombination all contributes to the shunt resistances of the device. The effect of series and resistance can be seen in Eq. (6) [33].

$$I = I_L - I_O\left[e^{\left\{\frac{q(M)}{AKT}\right\}} - 1\right] - \left(\frac{M}{R_{\text{shunt}}}\right) \tag{6}$$

where $M = V + I \times R_{\text{series}}$, I_L is the light-induced current, I_0 is the reverse saturation current of the diode, A is the ideality factor, K is the Boltzmann constant, T is

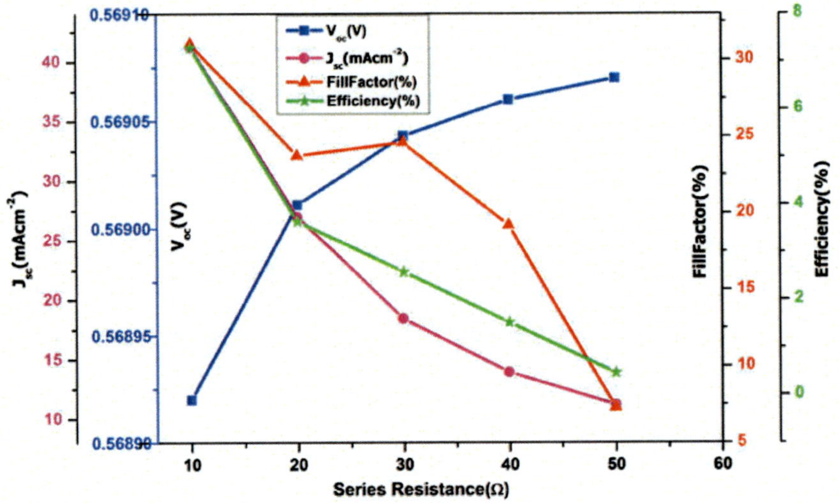

Fig. 5 Variation of photovoltaic parameters with change in series resistance

temperature, and q is the charge of the electron. Ideally, a zero series resistance and an infinite shunt resistance makes a perfect photovoltaic device. The series resistance and shunt resistance were both varied from 10 to 50 Ω. With the increase in series resistance, the FillFactor and short circuit current density declined rapidly to 11.1% and 7.3 mA cm^{-2} respectively, and the Voc remained almost constant at 0.5 V. The shunt resistance was also varied in the same range, but all the parameters improved drastically. Efficiency reached 14% at 50 Ω from 4% at 10 Ω. The short circuit density remained unchanged at 41 mA cm^{-2} as clearly indicated in Figs. 5 and 6.

3 Conclusions

In this work, a $Cs_2AuBiCl_6$ double perovskite-based solar cell was optimized with the SCAPS-1D simulation package. The work used a novel HTL material graphene oxide and commonly used ZnSe as ETL as a replacement to the widely used complex and costly charge carrier materials. The thickness of the absorber and the HTL material was optimized between 0.1 and 1 μm. The efficiency of the device increased and got saturated at 19% due to the between the carrier generation and bulk recombination within the absorber. Efficiency remained fairly constant with ETL thickness variation showing the minute role it plays in device working. The effects of the parasitic resistance were also thoroughly studied from 10 to 50 Ω. With the increase in series resistance, all other factors declined except the V_{oc} which remained constant at 0.56 V but with shunt resistance, the efficiency improved from 4 to 14% while the J_{sc} remained constant. Finally, the influence of operating temperature was studied, and

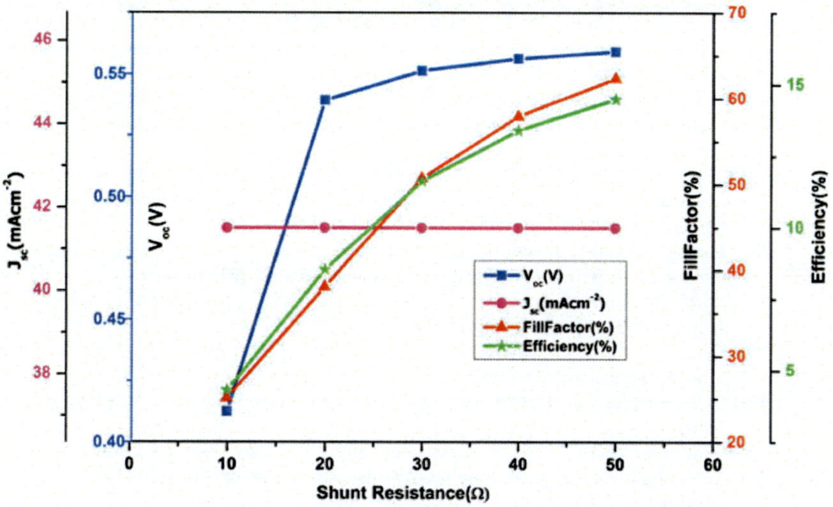

Fig. 6 Variation of photovoltaic parameters with change in shunt resistance

the optimum temperature for the device operation was found to be 300 K, any change from which adversely affects the device performance. In short, the $Cs_2AuBiCl_6$-based perovskite solar cells proved to be promising candidates in the journey to discover highly stable, non-toxic photovoltaic materials.

References

1. Kim JY, Lee J-W, Jung HS, Shin H, Park N-G (2020) High-efficiency perovskite solar cells. Chem Rev 120:7867–7918
2. Anon Light and oxygen induced degradation limits the operational stability of methylammonium lead triiodide perovskite solar cells. Energy Environ Sci (RSC Publishing)
3. Edri E, Kirmayer S, Cahen D, Hodes G (2013) High open-circuit voltage solar cells based on organic–inorganic lead bromide perovskite. J Phys Chem Lett 4:897–902
4. Xiao Z, Meng W, Wang J, Mitzi DB, Yan Y (2017) Searching for promising new perovskite-based photovoltaic absorbers: the importance of electronic dimensionality. Mater Horiz 4:206–216
5. Xiao Z, Du K-Z, Meng W, Wang J, Mitzi DB, Yan Y (2017) Intrinsic instability of $Cs_2In(I)M(III)X_6$ (M = Bi, Sb; X = halogen) double perovskites: a combined density functional theory and experimental study. J Am Chem Soc 139:6054–6057
6. Anon optical properties of $Cs_2NaBiCl_6$. Semantic Scholar
7. Morrs (1972) Anon crystal structure of $Cs_2NaBiCl_6$. Acta Crystallogr Sect B
8. Lozhkina OA, Murashkina AA, Elizarov MS, Shilovskikh VV, Zolotarev AA, Kapitonov YV, Kevorkyants R, Emeline AV, Miyasaka T (2017) Microstructural analysis and optical properties of $Cs_2BiAgBr_6$ halide double perovskite single crystals
9. Anon oxide perovskites, double perovskites and derivatives for electrocatalysis, photocatalysis, and photovoltaics. Energy Environ Sci (RSC Publishing)

10. Wei F, Deng Z, Sun S, Xie F, Kieslich G, Evans DM, Carpenter MA, Bristowe PD, Cheetham AK (2016) The synthesis, structure and electronic properties of a lead-free hybrid inorganic–organic double perovskite $(MA)_2KBiCl_6$ (MA = methylammonium). Mater Horiz 3:328–332
11. Zhao X-G, Yang J-H, Fu Y, Yang D, Xu Q, Yu L, Wei S-H, Zhang L (2017) Design of lead-free inorganic halide perovskites for solar cells via cation-transmutation. J Am Chem Soc 139:2630–2638
12. Guo T-M, Gao F-F, Li Z-G, Liu Y, Yu M-H, Li W (2020) Mechanical and acoustic properties of a hybrid organic–inorganic perovskite, TMCM-$CdCl_3$, with large piezoelectricity. APL Mater 8:101106
13. Anon $Cs_2AgInCl_6$ double perovskite single crystals: parity forbidden transitions and their application for sensitive and fast UV photodetectors. ACS Photonics
14. Anon vibrational analysis of the elpasolites Cs_2NaAlF_6 and Cs_2NaGaF_6 doped with Cr^{3+} ions by fluorescence spectroscopy. SpringerLink
15. Zhou (2018) Anon synthesis and photocatalytic application of stable lead-free $Cs_2AgBiBr_6$ perovskite nanocrystals. Small
16. Giustino F, Snaith HJ (2016) Toward lead-free perovskite solar cells. ACS Energy Lett 1:1233–1240
17. Goldschmidt VM (1926) Die Gesetze der Krystallochemie. Naturwissenschaften 14:477–485
18. Anon highly stable, phase pure $Cs_2AgBiBr_6$ double perovskite thin films for optoelectronic applications. J Mater Chem A (RSC Publishing)
19. Kale AJ, Chaurasiya R, Dixit A (2021) Inorganic lead-free $Cs_2AuBiCl_6$ perovskite absorber and Cu_2O hole transport material based single-junction solar cells with 22.18% power conversion efficiency. Adv Theory Simul 4:2000224
20. Rühle S (2016) Tabulated values of the Shockley–Queisser limit for single junction solar cells. Sol Energy 130:139–147
21. Kung P-K, Li M-H, Lin P-Y, Jhang J-Y, Pantaler M, Lupascu DC, Grancini G, Chen P (2020) Lead-free double perovskites for perovskite solar cells. Sol RRL 4:1900306
22. Rolston N, Printz AD, Hilt F, Hovish MQ, Brüning K, Tassone CJ, Dauskardt RH (2017) Improved stability and efficiency of perovskite solar cells with submicron flexible barrier films deposited in air. J Mater Chem A 5:22975–22983
23. Runa A, Feng S, Wen G, Feng F, Wang J, Liu L, Su P (2018) Highly reproducible perovskite solar cells based on solution coating from mixed solvents. J Mater Sci 53:3590–3603
24. Jeng (2013) Anon $CH_3NH_3PbI_3$ perovskite/fullerene planar-heterojunction hybrid solar cells. Adv Mater (Wiley Online Library)
25. Iqbal (2018) Anon novel graphene-based transparent electrodes for perovskite solar cells. Int J Energy Res (Wiley Online Library)
26. Milić (2018) Anon reduced graphene oxide as a stabilizing agent in perovskite solar cells. Adv Mater Interfaces (Wiley Online Library)
27. Anon a numerical study of high efficiency ultra-thin CdS/CIGS solar cells. Afr J Sci, Technol, Innov Dev 8(4)
28. Jayan K (2021) Anon comparative study on the performance of different lead-based and lead-free perovskite solar cells. Adv Theory Simul (Wiley Online Library)
29. Widianto E, Subama E, Nursam NM, Triyana K, Santoso I (2022) Design and simulation of perovskite solar cell using graphene oxide as hole transport material. AIP Conf Proc 2391:090011
30. Kale AJ, Chaurasiya R, Dixit A (2021) Inorganic lead-free $Cs_2AuBiCl_6$ perovskite absorber and Cu_2O hole transport material based single-junction solar cells with 22.18% power conversion efficiency. Adv Theory Simul 4:2000224

31. Srivastava A, Dua P, Lenka TR, Tripathy SK (2021) Numerical simulations on CZTS/CZTSe based solar cell with ZnSe as an alternative buffer layer using SCAPS-1D Mater Today: Proc 43:3735–3739
32. Khoshsirat N, Md Yunus NA, Hamidon MN, Shafie S, Amin N (2015) Analysis of absorber layer properties effect on CIGS solar cell performance using SCAPS. Optik 126:681–686
33. Karthick S, Velumani S, Bouclé J (2020) Experimental and SCAPS simulated formamidinium perovskite solar cells: a comparison of device performance. Sol Energy 205:349–357

Nanoscopic Pd(II)-Based Complexes with Poly-Ether Functionalized Ligand: The Crown Ether Analog

Debakanta Tripathy, Soumya Lipsa Rath, Srabani Srotwosini Mishra, and Dillip Kumar Chand

Abstract Mononuclear cis-protected Pd(II) complexes **1–4** with general formula $[Pd(N\text{-}N)L](NO_3)_2$ ((N-N) is cis-protecting units such as en, tmeda, bpy, and phen) were synthesized by combining pyridine appended poly-ether functionalized ligand (1,11-bis(4-pyridylcarboxy)-3,6,9-trioxaundecane) **L** with suitable palladium(II) components. These assemblies were characterized by NMR and ESI-MS. DFT calculation was undertaken to establish structure of these complexes. A clear concentration-dependent dynamic equilibrium in the solution state was observed. The existence of dynamic equilibrium was supplemented by variable concentration ^{1}H NMR study. DNA binding potential of these complexes was studied using molecular docking tool.

Keywords Palladium(II) · Self-assembly · Cis-protected · Poly-ether · Dynamic equilibrium · Crown ether

1 Introduction

Since the serendipitous discovery of crown ether by Pedersen, many research groups are working on developing synthetic molecular receptors for various applications. Metal-driven self-assembly provides an alternative to traditional organic synthesis for self-assembly of functional molecular receptor [1–14]. The beauty lies in the ease of its synthesis by exploiting the directional metal–ligand co-ordinate bond. The combination of metal component with suitably chosen ligand leads to the formation

D. Tripathy (✉) · S. S. Mishra · D. K. Chand
Department of Chemistry, School of Advanced Sciences, VIT-AP University, Amaravati, Andhra Pradesh 522237, India
e-mail: debakanta.tripathy@vitap.ac.in

D. Tripathy
Department of Chemistry, Indian Institute of Technology Madras, Chennai 600036, India

S. L. Rath
Department of Biotechnology, National Institute of Technology Warangal, Hanamkonda, Telangana 506004, India

N. M. Rao et al. (eds.), *Advanced Nanomaterials and Their Applications*,
Springer Proceedings in Materials 22, https://doi.org/10.1007/978-981-99-1616-0_4

of the targeted complex. The choice of building blocks and the reaction conditions play important role in deciding the geometry of final ensemble. Palladium(II) has been widely used as the metal precursor for the construction of self-assemblies [15].

Apart from the reaction conditions, the design of ligand is also crucial in construction of metal driven self-assembly. The structure and function of the self-assembled complex are greatly influenced by the nature of functional groups present on the ligand backbone. Pyridine-appended poly-ether ligands have already been used for preparation of metal-driven self-assembly [16–26]. It is observed that nature of the final ensemble formed largely depends on the flexibility of the ligand which again depends on the number of ethylene groups present on the ligand backbone. In most of the cases, self-assembly of these type of ligands with different metals form either a mononuclear or binuclear complex. However, in one case concentration-dependent formation of a mixture of macrocycle and single-stranded helix is observed [22]. Similarly, in another case the self-assembly process leading to the formation of a 2D-polycatenane is also reported [23]. Complexation property of the poly-ether ligand **L** (used in the present study, Fig. 1) with rhenium and ruthenium has been reported in the literature [19, 21, 25]. It produces mononuclear complexes when combined with rhenium [19, 21]; however, a mixture of mono- and binuclear complexes is formed upon combination with ruthenium [25]. The rhenium macrocycle of **L** shows cytotoxicity potential toward different cancer cell lines. Nevertheless, self-assembly of water-soluble metal complexes is always a challenging task. Herein we report the synthesis of a series of readily water soluble self-assembled Pd(II)-based macrocycles by employing the ligand **L** with different cis-protected Pd(II) components. A mixture of mononuclear and binuclear complexes which exist in concentration-dependent dynamic equilibrium (discussed in Sect. 3.3) is formed from the complexation of **L** with Pd(II) components. The DNA binding potential of these complexes is studied in silico using molecular docking tool, and the result shows strong binding ability of these complexes to DNA. The self-assembly of **L′** (3-pyridine analog of the ligand **L**) with palladium which produces only a series of mononuclear macrocycles has been reported in the literature [26]. Unlike **L**, the self-assembly of **L′** does not produce mixture of complexes in the solution state.

2 Experimental Section

2.1 Materials and Methods

$PdCl_2$ and isonicotinic acid, thionyl chloride, tetraethylene glycol, $AgNO_3$ along with all the common solvents were obtained from Aldrich and Spectrochem, India, respectively. All these materials were used as such without any purification unless specified. All the deuterated solvents were procured either from Aldrich or Cambridge Isotope Laboratories. ^{1}H and ^{13}C NMR spectral data were obtained from Bruker 400 MHz FT NMR or Bruker 500 MHz FT NMR spectrometer in $CDCl_3$ and D_2O (by using TMS

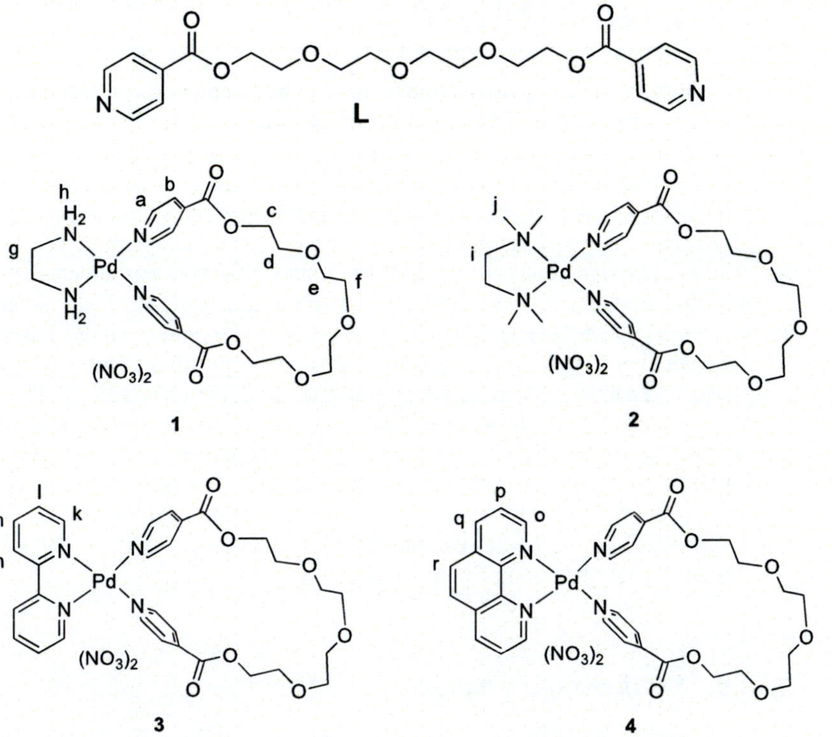

Fig. 1 Structure of the ligand **L** and complexes cis-[Pd(en)**L**]$(NO_3)_2$, **1**; cis-[Pd(tmeda)**L**]$(NO_3)_2$, **2**; cis-[Pd(bpy)**L**]$(NO_3)_2$, **3** and cis-[Pd(phen)**L**]$(NO_3)_2$, **4** (with proton numbering)

in $CDCl_3$ as an external reference). The ESI mass spectra were obtained from a Micromass Q-TOF mass spectrometer. The cis-protected Pd(II) components were obtained following conventional processes. The ligand **L** was synthesized by following the literature procedure and is described in Sect. 2.3 [26].

2.2 Preparation of Isonicotinoyl Chloride Hydrochloride

To a 100 mL two naked round bottom flask, isonicotinic acid (0.732 g, 0.595 mol) was added followed by quick addition of thionyl chloride (6 mL) at room temperature. Then the reaction mixture was refluxed at 80 °C for about 30 min. The clear solution obtained was evaporated to dryness under reduced pressure to get the desired product as white crystalline solid (m.p. 152 °C). The isonicotinoyl chloride hydrochloride obtained was immediately used for the preparation of ligand **L**.

2.3 Synthesis of Ligand L

0.758 g (0.426 mol) of isonicotinoyl chloride hydrochloride was taken in a 100 mL two naked round bottom flask, and 15 mL of dichloromethane was added to it under nitrogen atmosphere. The suspension was stirred thoroughly for 15 min and then 0.20 mL (0.222 g, 0.210 mol) of tetraethylene glycol was added. To this reaction mixture kept in an ice bath 1 mL of triethylamine was added drop-wise over a period of 30 min. The resulting mixture was stirred at room temperature for 24 h under nitrogen atmosphere. Then $NaHCO_3$ solution (10% w/v) was added slowly to the resulting mixture until the cessation CO_2 evolution. The organic layer was separated, washed with distilled water and dried over anhydrous sodium sulfate. Ligand **L** was obtained as pale-yellow solid after complete removal of the solvent under reduced pressure. Yield = 0.611 g (72%). m.p. 66 °C. ^{1}H NMR (400 MHz, $CDCl_3$, 298 K): δ = 8.78 (dd, 4H, J1 = 4.4, J2 = 4.4 Hz, H_a), 7.86 (dd, 4H, J1 = 4.8, J2 = 4.8 Hz, H_b), 4.52–4.49 (m, 4H, H_c), 3.84–3.82 (m, 4H, H_d), 3.69 (m, 8H, H_e and H_f) ppm. ^{1}H NMR (400 MHz, D_2O, 25 °C): δ = 8.60 (d, 4H, J = 5.6 Hz, H_a), 7.87 (d, 4H, J = 6.0 Hz, H_b), 4.50 (t, 4H, J = 4.4 Hz, H_c), 3.89 (t, 4H, J = 4.4 Hz, H_d), 3.75–3.71 (m, 8H, H_e and H_f). ^{13}C NMR (100 MHz, $CDCl_3$, 25 °C): δ = 165.18, 150.69, 137.37, 123.02, 70.78, 69.07, 64.88 ppm.

2.4 Synthesis of Complex 1

To the solution of [Pd(en)$(NO_3)_2$] (58 mg, 0.2 mmol in 5 mL of CH_3CN) was added the ligand **L** (81 mg, 0.2 mmol), then the reaction mixture was stirred at room temperature for about 5 h. The colorless solution obtained was centrifuged, decanted to a watch glass and left for drying in open air followed by drying under reduced pressure. Complex **1** was obtained as a brown color solid which turns to sticky mass upon exposure to air. Yield 133 mg (96%). ^{1}H NMR (400 MHz, D_2O, 298 K): δ = 8.84 (dd, 4H, J1 = 5.2 Hz, J2 = 5.2 Hz, H_a), 8.00 (dd, 4H, J1 = 5.2, J2 = 5.2 Hz, H_b), 4.55–4.53 (m, 4H, H_c), 3.87–3.85 (m, 4H, H_d), 3.67–3.65 (m, 4H, H_e), 3.59–3.57 (m, 4H, H_f), 2.95 (s, 4H, H_g) ppm. ^{13}C NMR (100 MHz, D_2O, 298 K): δ = 164.83, 152.33, 141.10, 126.19, 70.11, 69.91, 68.70, 65.88, 47.04 ppm. ESI-MS: m/z: 285 for [**1**-2(NO_3)]$^{2+}$.

2.5 Synthesis of Complex 2

Complex **2** was prepared following very similar method to that of complex **1** using equimolar amount of corresponding Pd(II) component, [Pd(tmeda)$(NO_3)_2$] (52 mg, 0.150 mmol), and ligand **L** (61 mg, 0.150 mmol). Yield = 105 mg (93%). ^{1}H NMR (400 MHz, D_2O, 298 K): δ = 9.17 (dd, 4H, J1 = 5.6 Hz, J2 = 5.2 Hz, H_a), 8.04

(dd, 4H, J1 = 5.2, J2 = 5.2 Hz, H_b), 4.55–4.53 (m, 4H, H_c), 3.87–3.85 (m, 4H, H_d), 3.69–3.67 (m, 4H, H_e), 3.61–3.59 (m, 4H, H_f), 3.18 (s, 4H, H_i), 2.82 (s, 4H, H_j) ppm. ^{13}C NMR (100 MHz, D_2O, 298 K): δ = 164.50, 152.0, 141.17, 126.81, 70.07, 69.77, 68.62, 65.75, 63.03, 50.67 ppm. ESI-MS: m/z: 688, and 313 for [**2**-NO_3)]$^+$ and [**2**-2(NO_3)]$^{2+}$ respectively.

2.6 *Synthesis of Complex 3*

Complex **3** was prepared following very similar method to that of complex **1** using equimolar amount of corresponding Pd(II) component, [Pd(bpy)(NO_3)$_2$] (55 mg, 0.142 mmol) and ligand **L** (58 mg, 0.142 mmol). Yield = 107 mg (95%). ^{1}H NMR (400 MHz, D_2O, 298 K): δ = 9.25 (dd, 4H, J1 = 5.2 Hz, J2 = 5.2 Hz, H_a), 8.51 (d, 2H, J = 7.6 Hz, H_n), 8.42 (dd, 2H, J1 = 7.6, J2 = 8.0 Hz, H_m), 8.17 (dd, 4H, J1 = 5.2, J2 = 5.2 Hz, H_b), 7.88 (d, 2H, J = 4.8 Hz, H_k), 7.66 (m, 2H, H_l), 4.61–4.59 (m, 4H, H_c), 3.91–3.89 (m, 4H, H_d), 3.72–3.70 (m, 4H, H_e), 3.63–3.61 (m, 4H, H_f) ppm. ^{13}C NMR (100 MHz, D_2O, 298 K): δ = 164.58, 156.83, 152.54, 150.66, 143.05, 141.82, 128.30, 127.17, 124.56, 70.11, 69.88, 68.69, 65.94 ppm. ESI-MS: m/z: 332 for [**3**-2(NO_3)]$^{2+}$.

2.7 *Synthesis of Complex 4*

Complex **4** was prepared following very similar method to that of complex **1** using equimolar amount of corresponding Pd(II) component, [Pd(phen)(NO_3)$_2$] (52 mg, 0.127 mmol) and ligand **L** (52 mg, 0.127 mmol). Yield = 101 mg (97%). ^{1}H NMR (400 MHz, D_2O, 298 K): δ = 9.25 (d, 4H, J = 6.4 Hz, H_a), 8.92 (d, 2H, J = 8.0 Hz, H_q), 8.24 (s, 2H, H_r), 8.122 (d, 4H, J = 5.2 Hz, H_o), 8.16 (d, 4H, J = 6.4 Hz, H_b), 7.90 (dd, J1 = 5.6, J2 = 8.4 Hz, 2H, H_p), 4.57–4.55 (m, 4H, H_c), 3.87–3.85 (m, 4H, H_d), 3.67–3.65 (m, 4H, H_e), 3.59–3.57 (m, 4H, H_f) ppm. ^{13}C NMR (100 MHz, D_2O, 298 K): δ = 164.63, 152.79, 151.25, 147.22, 142.00, 141.93, 131.44, 128.32, 127.20, 126.34, 70.12, 69.90, 68.70, 65.97 ppm. ESI-MS: m/z: 345 for [**4**-2(NO_3)]$^{2+}$.

3 Results and Discussion

3.1 *Synthesis and Characterization of Complexes 1–4*

Complexes **1**–**4** were obtained in near quantitative yield by equimolar combination of ligand **L** with corresponding cis-Pd(II) in acetonitrile (Fig. 1). All these complexes were characterized by ^{1}H (Fig. 2) and ^{13}C NMR spectroscopy. All the peaks were thoroughly assigned by using 2D NMR techniques such as H–H COSY and C–H COSY (see supplementary information). A single set of sharp peaks obtained in NMR spectrum at 5 mM concentration of palladium(II) indicates the formation of a single compound in the solution. Figure 2 shows the comparison of ^{1}H NMR for the ligand **L** with complexes **1–4**. The signature downfield shift of the pyridine α proton of the ligand in all the complexes indicates the withdrawal of electron density upon complexation with palladium unit. The downfield shift is marginal in case of complex **1**; however, it is significant in **2–4**. Pyridine β proton shows minimal downfield shift due to its distant position from the complexation site. Similarly, the aliphatic protons are less influenced upon metal coordination owing to their remote location to the palladium unit. H_c shows marginal downfield shift however, H_d–H_f are slightly up-field shifted. The chemical shift change values for various protons are summarized in Table 1. Similarly, the signals due to protons present in the cis-protecting units exhibit significant complexation-induced downfield shift. ESI-MS data for complexes **1–4** show the 1:1 stoichiometry of metal–ligand self-assembly. The peak at m/z = 285.09 corresponds to the molecular fragment [**1**-2NO$_3$]$^{2+}$, and expansion of the bundle of peaks show the correct isotopic pattern for the said fragment. Similarly, peaks at 313.05, 688.18, 332.94, and 345.07 corresponds to molecular fragments [**2**-2NO$_3$]$^{2+}$, [**2**-NO$_3$]$^{+}$, [**3**-2NO$_3$]$^{2+}$, and [**4**-2NO$_3$]$^{2+}$ respectively. The envelope of peaks obtained is in good agreement with the theoretical pattern for above-mentioned fragments (see supplementary information). Molecular structure of these complexes was obtained by DFT calculation which is discussed in following section.

3.2 DFT Calculations for Complexes 1–4

After several futile attempts to grow single crystals for X-ray analysis, DFT calculation was carried out to get the geometry optimized structure of complexes **1–4**. DFT study for these complexes was performed with the Gaussian 16 software package. B3LYP (Becke's three parameter hybrid functional using the LYP correlation) functional with 6-31G* basis set for C, N, O and H atoms and LANL2DZ for Pd atom was used for optimizing molecule geometry. After geometry was optimized, the frequencies were calculated on the structures in order to confirm the absence of any imaginary frequencies. The geometry optimized structure of **1–4** obtained from DFT calculations are shown in Fig. 3. Palladium describes the square planar geometry and coordinated to one cis-protecting unit as well as one unit of ligand **L** in a chelating fashion to form the mononuclear loop. Bond length and bond angles around palladium(II) for complexes **1–4** are comparable to the palladium complexes documented in the literature [23, 24, 26–28]. Major bond lengths and bond angles for complexes **1–4** are listed in Table 2. It was observed from the energy minimized structures that complexes **1–4** possess well defined cavity which contain multiple oxygen atoms very similar to that of a crown ether and hence termed as metallo-crown ethers.

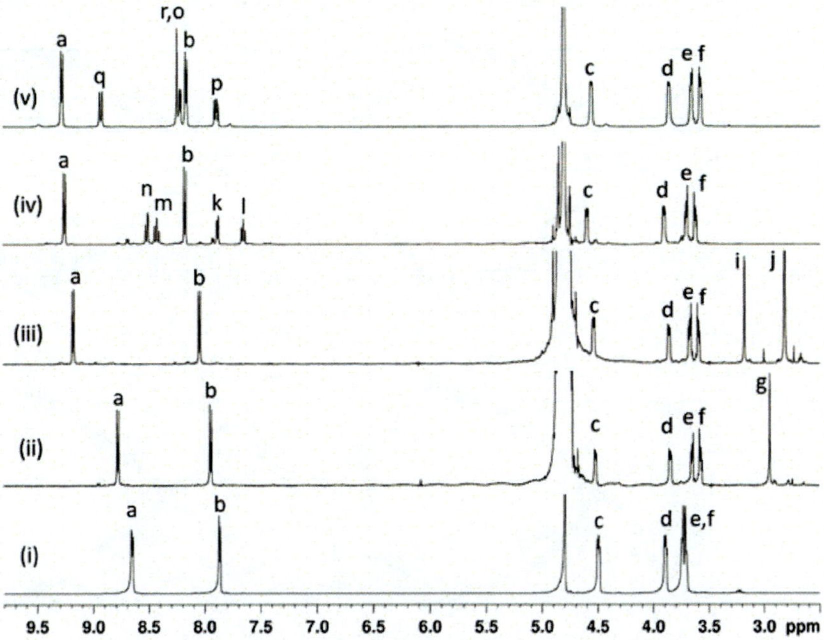

Fig. 2 400 MHz ^{1}H NMR in D_2O for (i) the ligand **L**, (ii–v) complexes; (ii) **1**, (iii) **2**, (iv) **3**, and (v) **4** (at 5 mM concentration of Pd(II), for proton numbering refer to Fig. 1)

Table 1 Change in chemical shift values for various protons of **L** in complexes **1–4**

Complexes	Change in chemical shift values ($\Delta\delta$ in ppm)					
	H_a	H_b	H_c	H_d	H_e	H_f
1	0.18	0.13	0.04	−0.03	−0.08	−0.13
2	0.52	0.17	0.04	−0.03	−0.06	−0.11
3	0.59	0.64	0.10	0.01	−0.03	−0.09
4	0.61	1.05	0.06	−0.03	−0.08	−0.14

3.3 Dynamic Behavior of Complexes 1–4 in Solution State

Existence of dynamic equilibrium between self-assemblies of varying nuclearity is a well-known process in metal driven self-assembly [25–32]. Complexes **1–4** exist in dynamic equilibrium with their corresponding higher homologue (**1′–4′**) in the solution state (Fig. 4). The dynamic process was probed by recording ^{1}H NMR over a wide range of concentration. The ^{1}H NMR recorded for complexes **1–4** in the concentration range 5–40 mM (Figs. 5, 6, 7 and Fig. 8) with respect to Pd(II) show

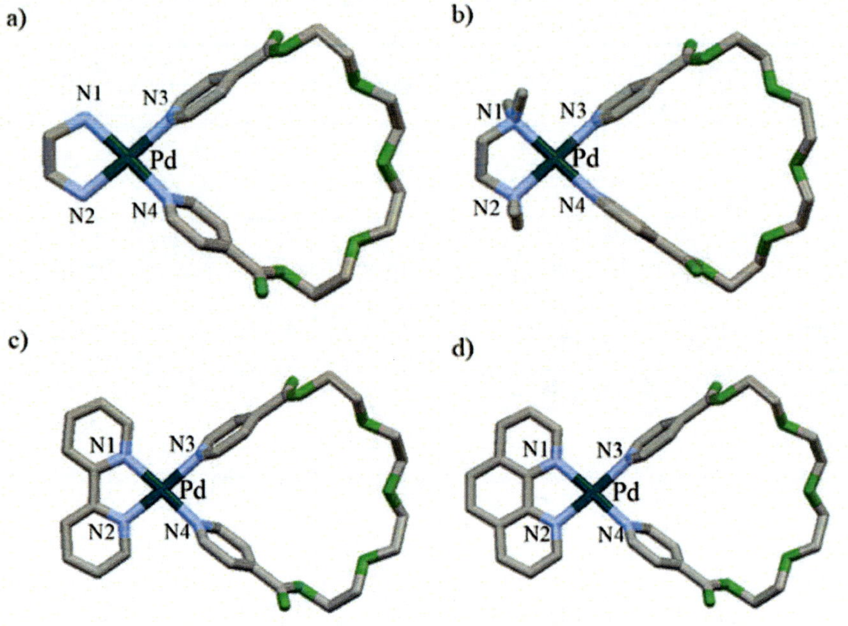

Fig. 3 Energy-optimized structures of complexes **1–4**; **a) 1, b) 2, c) 3,** and **d) 4** (hydrogen atoms and anions are not shown, perspective view of complexes perpendicular to the PdN_4 plane is shown)

Table 2 Various Pd–N bond lengths and bond angles around Pd units in complexes **1–4**

Complexes	Pd–N1 (Å)	Pd–N2 (Å)	Pd–N3 (Å)	Pd–N4 (Å)	N1PdN2 (°)	N2PdN4 (°)	N4PdN3 (°)	N3PdN1 (°)
1	2.112	2.112	2.069	2.070	82.38	95.06	87.59	94.97
2	2.137	2.138	2.095	2.095	85.01	94.94	85.27	94.83
3	2.085	2.085	2.080	2.080	80.03	97.28	85.70	97.05
4	2.074	2.075	2.077	2.078	80.89	96.46	86.42	96.28

two set of peaks indicating the presence of a concentration dependent dynamic equilibrium between complexes **1–4** and their corresponding higher homologue which is supposed to be the binuclear species. As can be seen from the NMR, at 5 mM concentration with respect to Pd(II), a single species exist in the solution state and with gradual increase in concentration leads to the formation of a new species. At around 20 mM concentration, a mixture two different self-assemblies can be clearly seen. It can also be observed from the variable concentration NMR (Fig. 5) that the percentage of the corresponding binuclear species for complex **1** at 40 mM is higher than that of complexes **2–4** at similar concentration. However, no such dynamic behavior was reported in solution state for corresponding Pd(II) complexes of **L′** (a 3-pyridyl analog of ligand **L**) [22].

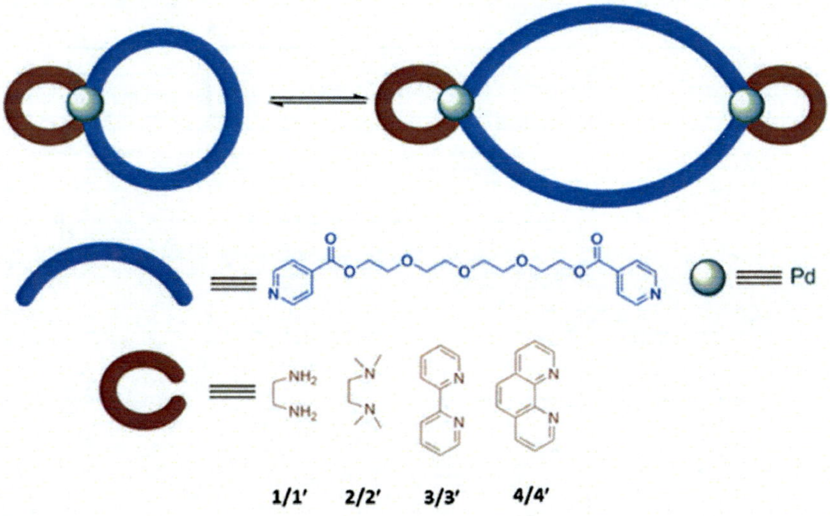

Fig. 4 Cartoon representation of dynamic equilibrium between complexes **1–4** and their corresponding higher homologue **1′–4′** (only cationic part is shown)

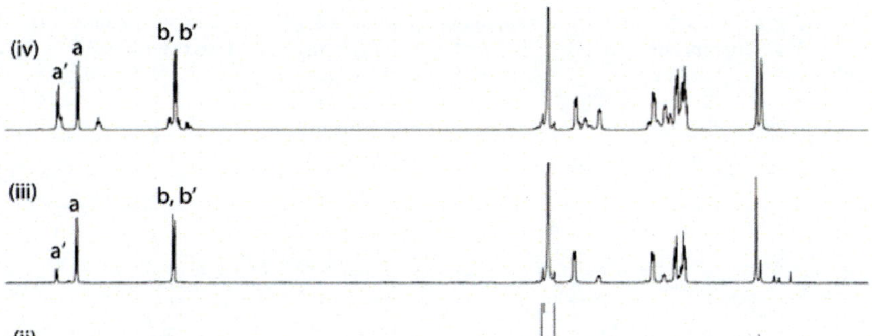

Fig. 5 400 MHz ^{1}H NMR for complexes $[Pd(en)\mathbf{L}](NO_3)_2$, **1** and $[Pd_2(en)_2(\mathbf{L})_2](NO_3)_4$, **1′** at different concentrations; (i) 5 mM, (ii) 10 mM, (iii) 20 mM, and (iv) 40 mM

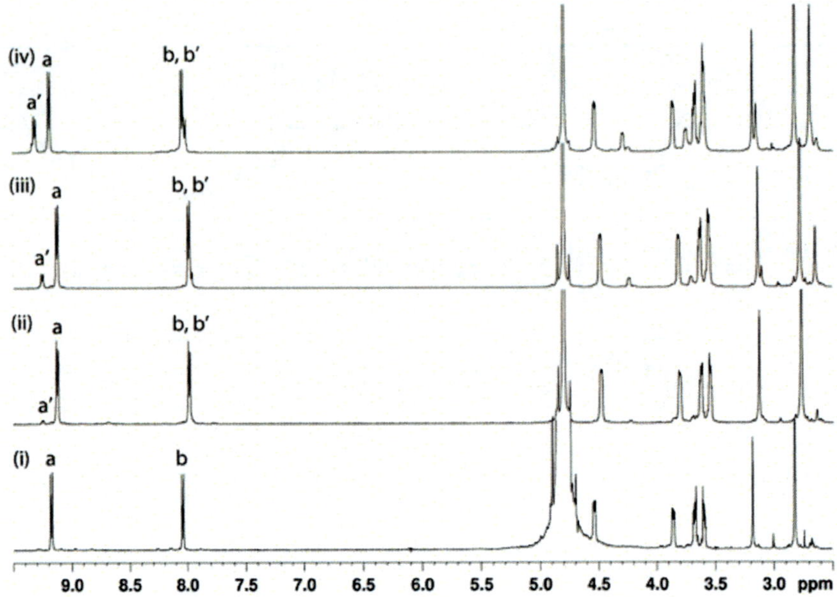

Fig. 6 400 MHz ^{1}H NMR for complexes [Pd(tmeda)**L**]$(NO_3)_2$, **2** and $[Pd_2(tmeda)_2(\mathbf{L})_2](NO_3)_4$, **2′** at different concentrations; (i) 5 mM, (ii) 10 mM, (iii) 20 mM, and (iv) 40 mM

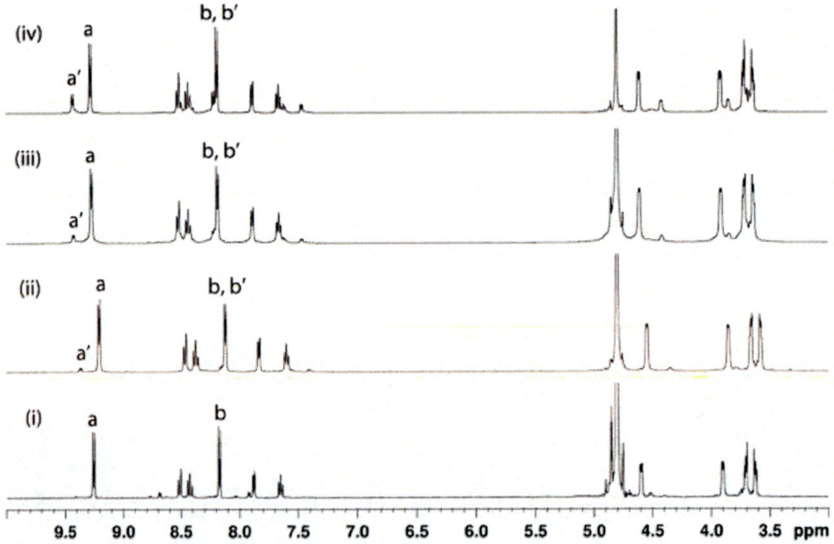

Fig. 7 400 MHz ^{1}H NMR for complexes [Pd(bpy)**L**]$(NO_3)_2$, **3** and $[Pd_2(bpy)_2(\mathbf{L})_2](NO_3)_4$, **3′** at different concentrations; (i) 5 mM, (ii) 10 mM, (iii) 20 mM, and (iv) 40 mM

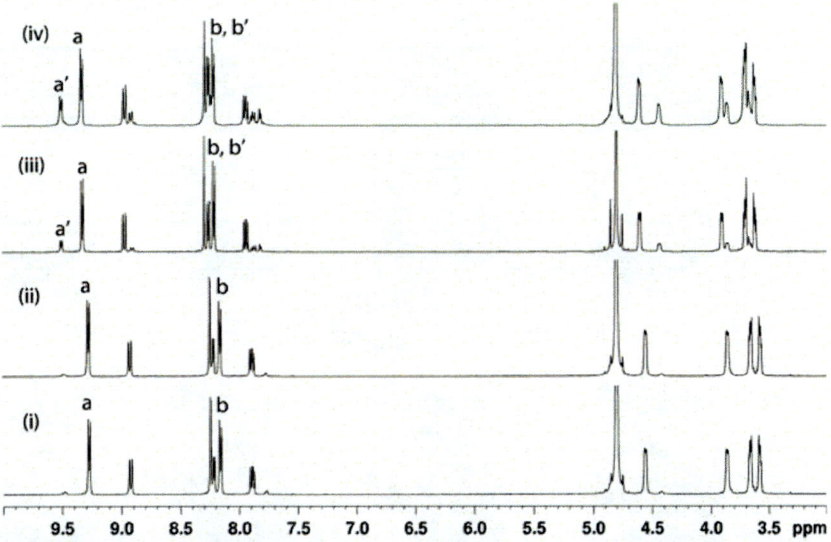

Fig. 8 400 MHz ^{1}H NMR for complexes [Pd(phen)**L**]$(NO_3)_2$, **4** and $[Pd_2(phen)_2(\mathbf{L})_2](NO_3)_4$, **4′** at different concentrations; (i) 5 mM, (ii) 10 mM, (iii) 20 mM, and (iv) 40 mM

3.4 Docking Studies of Complexes 1–4 with B-DNA

Molecular docking is an important tool to calculate the interaction energy of metal complexes with the biomolecules [29–31, 33–35]. Molecular docking of the compounds with B-DNA gave the energy values for complexes **1–4** as −9.45, −7.16, −8.19, and −8.46 kcal/mol, respectively. According to the results, the compound **1** has the maximum binding affinity and **2** has the least. To gain further information on the mode of binding we visualized the docked complexes using PyMOL [32, 36]. Figure 9a shows the docked complex of **1** with DNA, where the compound was found to bind in the major groove of the DNA. However, in docked DNA complexes with **2**, **3**, and **4** (Fig. 9b–d), the molecules were found to bind at the DNA minor groove. Specifically, the orientation and position of **2** with the DNA appeared distorted when compared to **3** and **4** where the ligand perfectly rests in the minor groove. The similarity in the orientation of **3** and **4** also explains the similar docking energies of the complexes. As seen in Fig. 10a, **1** makes large numbers of salt bridge interactions with the DNA, **2** shows a bifurcated salt-bridge from carbonyl oxygen to DNA NH group and one of the oxygen atoms of the phosphate group, respectively. The binding of both **3** and **4** was largely similar with two salt-bridge interactions between one of the ether oxygen and backbone NH group; carbonyl oxygen and backbone oxygen atoms of phosphate group, respectively. The slight difference in the docking energy could be due to minor structural change between both the molecules.

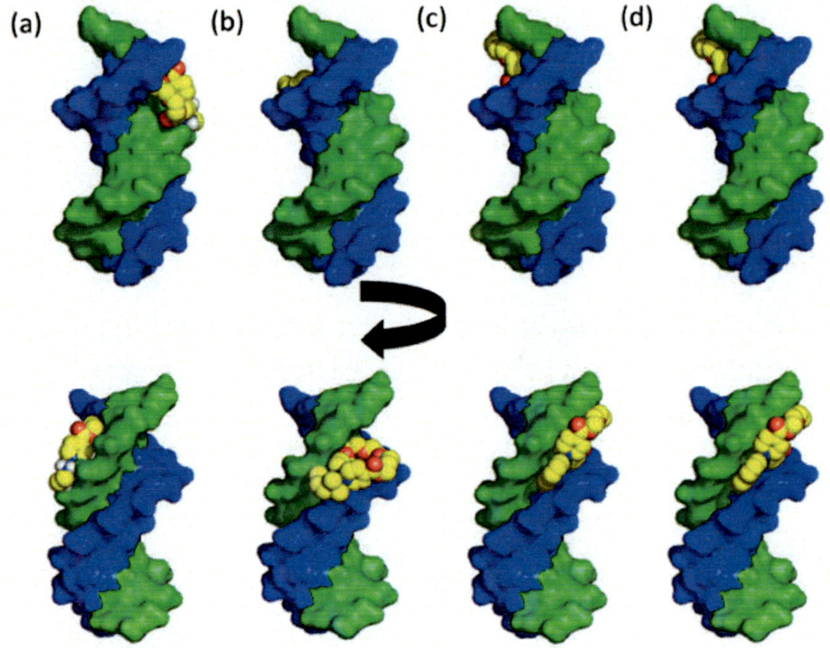

Fig. 9 DNA-molecule docked complexes showing the orientation of molecules (top: front view and bottom: back view). The complexes with the lowest docking energies are shown for **a** **1**, **b** **2**, **c** **3**, and **d** **4**. The molecules are shown as balls and colored in CPK; DNA backbones are shown in green and blue colors in surface mode

4 Conclusions

In conclusion, four water soluble mononuclear Pd(II) complexes **1–4** has been synthesized and characterized by NMR, ESI-MS, and DFT calculations. It is observed that complexes **1–4** exist in dynamic equilibrium with their corresponding binuclear species in the solution state. The ratio of mononuclear and binuclear complexes can be varied by varying the overall concentration. The binding energy for complexes **1–4** with B-DNA is found to be −9.45, −7.16, −8.19, and −8.46 kcal/mol, respectively.

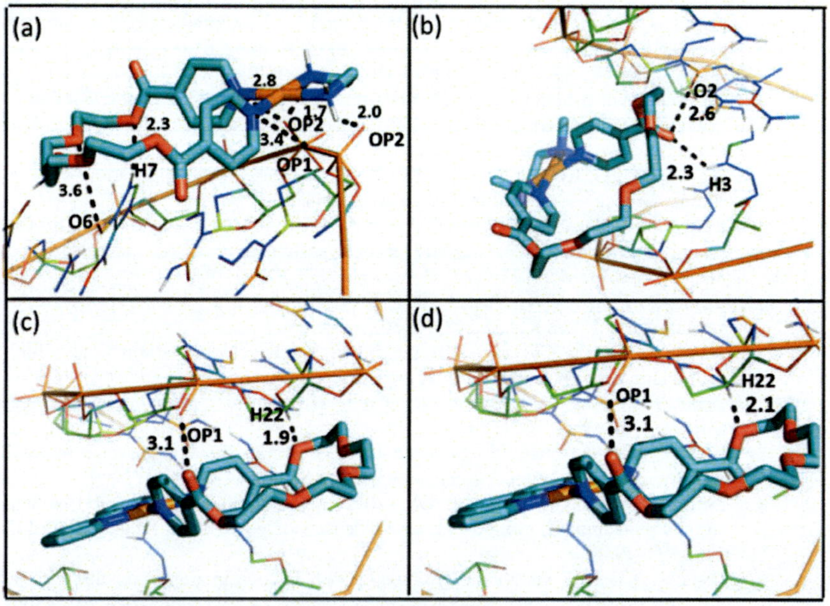

Fig. 10 Detailed molecular interaction of molecules (in licorice) with DNA (shown as sticks) for complexes **a 1**, **b 2**, **c 3**, and **d 4**. All atoms are colored in CPK

Acknowledgements DT thanks CSIR for a fellowship. DT, SSM, and DKC thank IIT Madras for the infrastructure facility. SLR thanks NITW for research seed grant P1131.

Declarations The authors have no competing interests to declare that are relevant to the content of this article.

Supplementary Data Supplementary data (NMR and ESI MS) associated with this article can be found, in the online version.

References

1. Pullen S, Tessarolo J, Clever GH (2021) Increasing structural and functional complexity in self-assembled coordination cages. Chem Sci 12:7269–7293
2. Debata NB, Tripathy D, Sahoo HS (2019) Development of coordination driven self-assembled discrete spherical ensembles. Coord Chem Rev 387:273–298
3. Roberts DA, Pilgrim BS, Nitschke JR (2018) Covalent post-assembly modification in metallosupramolecular chemistry. Chem Soc Rev 47:626–644
4. Pullen S, Clever GH (2018) Mixed-ligand metal-organic frameworks and heteroleptic coordination cages as multifunctional scaffolds—a comparison. Acc Chem Res 51:3052–3064
5. Lewis JEM, Beer PD, Loeb SJ, Goldup SM (2017) Metal ions in the synthesis of interlocked molecules and materials. Chem Soc Rev 46:2577–2591

6. Chen L-J, Yang H-B, Shionoya M (2017) Chiral metallosupramolecular architectures. Chem Soc Rev 46:2555–2576
7. He Z, Jiang W, Schalley CA (2015) Integrative self-sorting: a versatile strategy for the construction of complex supramolecular architecture. Chem Soc Rev 44:779–789
8. Yoshizawa M, Klosterman JK (2014) Molecular architectures of multi-anthracene assemblies. Chem Soc Rev 43:1885–1898
9. Percástegui EG, Ronson TK, Nitschke JR (2020) Design and applications of water-soluble coordination cages. Chem Rev 120:13480–13544
10. Fang Y, Powell JA, Li E, Wang Q, Perry Z, Kirchon A, Yang X, Xiao Z, Zhu C, Zhang L, Huang F, Zhou HC (2019) Catalytic reactions within the cavity of coordination cages. Chem Soc Rev 48:4707–4730
11. Zhang D, Ronson TK, Nitschke JR (2018) Functional capsules via subcomponent self-assembly. Acc Chem Res 51:2423–2436
12. Wang W, Wanga YX, Yang HB (2016) Supramolecular transformations within discrete coordination-driven supramolecular architectures. Chem Soc Rev 45:2656–2693
13. Galan A, Ballester P (2016) Stabilization of reactive species by supramolecular encapsulation. Chem Soc Rev 45:1720–1737
14. Ahmad N, Younus HA, Chughtai AH, Verpoort F (2015) Metal–organic molecular cages: applications of biochemical implications. Chem Soc Rev 44:9–25
15. Debata NB, Tripathy D, Chand DK (2012) Self-assembled coordination complexes from various palladium(II) components and bidentate or polydentate ligands. Coord Chem Rev 256:1831–1945
16. Tripathy D, Debata NB, Naik KC, Sahoo HS (2022) Coordination driven discrete metallopolygons and cages from unsymmetric bidentate ligands. Coord Chem Rev 456:214396
17. Dürr H, Kilburg H, Bossmann S (1990) Efficient synthesis of crown ester-linked ruthenium complexes. A new class of a supramolecular sensitizer-crown ester assembly for electron transfer. Synthesis 773–778
18. Ramakrishna B, Ashok Kumar C, Logesh TJ, Manimaran B (2017) Oxamidato pillared hetero-ligated dirhenium(I) metallacrown ethers: synthesis, spectroscopic and structural characterization. J Organomet Chem 828:116–121
19. Bélanger S, Gilbertson M, Yoon DI, Stern CL, Dang X, Hupp JT (1999) A luminescent tricarbonylchlororhenium(I) complex featuring a flexible "crown ether" ligand. Manipulation of photoexcited state properties via binding of small cations. J Chem Soc Dalton Trans 3407–3412
20. Chevrier I, Sagué JL, Brunetto PS, Khanna N, Rajacic Z, Fromm KM (2013) Rings, chains and helices: new antimicrobial silver coordination compounds with (iso-)nicotinic acid derivatives. Dalton Trans 42:217–231
21. Ashok Kumar C, Nagarajaprakash R, Victoria W, Veena V, Sakthivel N, Manimaran B (2016) Synthesis, characterisation and cytotoxicity studies of manganese(I) and rhenium(I) based metallacrown ethers. Inorg Chem Commun 64:39–44
22. Fromm KM, Sagué Doimeadios JL, Robin AY (2005) Concomitant crystallization of two polymorphs—a ring and a helix: concentration effect on supramolecular isomerism. Chem Commun 4548–4550
23. Sagué JL, Fromm KM (2006) The first two-dimensional polycatenane: a new type of robust network obtained by Ag-connected one-dimensional polycatenanes. Cryst Growth Des 6:1566–1568
24. Song LC, Zhang WX, Hu QM, Zhao LQ (2004) Synthesis and characterization of a novel bicyclic dimer $[CuBr(pcadgd)]_2$ formed via self-assembly of 4-pyridinecarboxylic acid diethylene glycol diester (pcadgd) with copper(I) bromide. Transit Met Chem 29:603–606
25. Dürr H, Kilburg H, Bossmann S (1990) Efficient synthesis of crown ester-linked ruthenium complexes. A new class of a supramolecular sensitizer-crown ester assembly for electron transfer. Synthesis 9:773–778
26. Sahoo HS, Tripathy D, Chakrabortty S, Bhat S, Kumbhar A, Chand DK (2013) Self-assembled mononuclear palladium(II) based molecular loops. Inorg Chim Acta 400:42–50

27. Ganta S, Jagan R, Chand DK (2019) Crystal engineering with palladium(II)-based self-assembled binuclear complexes as tectons. Inorg Chim Acta 484:33–41
28. Tripathy D, Ramkumar V, Chand DK (2013) Toppled molecular-domino sets by self-assembly of self-assembly: the π-polymers. Cryst Growth Des 13:3763–3772
29. Sautter A, Schmid DG, Jung G, Würthner F (2001) A triangle–square equilibrium of metallosupramolecular assemblies based on Pd(II) and Pt(II) corners and diazadibenzoperylene bridging ligands. J Am Chem Soc 123:5424–5430
30. Ferrer M, Mounir M, Rossell O, Ruiz E, Maestro MA (2003) Equilibria between metallosupramolecular squares and triangles with the new rigid linker 1,4-bis(4-pyridyl)tetrafluorobenzene. Experimental and theoretical study of the structural dependence of NMR data. Inorg Chem 42:5890–5899
31. Uehara K, Kasai K, Mizuno N (2007) Syntheses and characterizations of palladium-based molecular triangle/square compounds and hybrid composites with polyoxometalates. Inorg Chem 46:2563–2570
32. Debata NB, Tripathy D, Ramkumar V, Chand DK (2010) Coordination-driven self-assembly in a single pot. Tetrahedron Lett 51:4449–4451
33. Tripathy D, Pal AK, Rath SL, Panda BB, Hanan GS, Chand DK (2021) Synthesis, characterization and molecular docking study of nitro(4′-(2-pyridyl)-2,2′:6′,2″-terpyridyl) palladium(II) nitrate. Inorg Chem Commun 126:108494
34. Ekennia AC, Osowole AA, Olasunkanmi LO, Onwudiwe DC, Olubiyi OO, Ebenso EE (2017) Synthesis, characterization, DFT calculations and molecular docking studies of metal (II) complexes. J Mol Struct 1150:279–292
35. El-Deen IM, Shoair AF, El-Bindary MA (2018) Synthesis, structural characterization, molecular docking and DNA binding studies of copper complexes. J Mol Liq 249:533–545
36. The PyMOL molecular graphics system. Version 1.2r3pre. Schrödinger, LLC

Preparation of Hydrotalcite–$CdPS_3$ Hybrid Solid from the Exfoliated Inorganic Nanosheets

Rajesh Chalasani

Abstract The delamination of layered solids, hydrotalcite (layered double hydroxides-LDH) clay, and cadmium thiophosphates ($CdPS_3$), has been achieved by intercalating a surfactant followed by dispersing in a nonpolar solvent. A anionic surfactant dodecyl sulfate (DDS) and the cationic surfactant dioctadecyldimethylammonium bromide (DODMA) have been used for the intercalation of LDH and $CdPS_3$, respectively. In dispersion, LDH and $CdPS_3$ nanosheets are electrically neutral as surfactant chains remain tethered to the layers. It is shown that the 1:1 mixtures of dispersions of $CdPS_3$–DODMA with LDH-DDS nanosheets can self-assemble, on solvent evaporation, to give a new layered solid with periodically alternating $CdPS_3$ and LDH layers. The driving force for the self-assembly to form periodic alternating layered structure could be the attractive force between the neutral exfoliated nanosheets.

Keywords Self-assembly · Organic–inorganic nanostructures · Two-dimensional materials · Solid-state chemistry

1 Introduction

Synthesis of ordered superstructures by self-directed assembly from the basic structural motif remains a challenge in solid state chemistry. Crystal engineering, uses synthetic building blocks, has been successfully implemented in organic solid state and framework solids for designing new solids [1]. The method of deconstructing a solid into basic structural block and reconstruct a new solid from them is yet to be stabilized for other crystalline solids. The idea of deconstruction, like exfoliation and delamination, followed by reconstruction can be successful in preparing new layered hybrids as the delamination procedures are reasonably well established [2–6].

The designing of functional nanocomposites and nanostructures from highly dispersed monolayers, which are obtained from the delamination of inorganic layered

R. Chalasani (✉)
Department of Chemistry, School of Advanced Sciences, VIT-AP University, Amaravati, Andhra Pradesh 522237, India
e-mail: rajesh.c@vitap.ac.in

N. M. Rao et al. (eds.), *Advanced Nanomaterials and Their Applications*,
Springer Proceedings in Materials 22, https://doi.org/10.1007/978-981-99-1616-0_5

solids, has attracted considerable interest in recent times [7–11]. The synthesis of new layered hybrids has been attempted by combining the dispersions of two different layered materials by layer-by-layer method [12–19]. In layer-by-layer method, the hybrids are prepared by depositing a oppositely charged species on a suitable substrate. The simplest way of preparing an ordered layered structure would be the mixing of dispersions of exfoliated inorganic nanosheets and allow them to self-assemble.

In this study, a hybrid solid has been synthesized by mixing the delaminated dispersions of neutral $CdPS_3$ and the anionic clay Mg–Al LDH. The structure of the $CdPS_3$ phase may be viewed as built from blocks of CdS_6 and P_2S_6 polyhedra which are linked by edge-sharing to form sheets. The $CdPS_3$ sheets stack one upon another and form 3D lattice. $CdPS_3$ is better represented as a thiophosphate salt, $Cd_2{}^{4+}(P_2S_6)^{4-}$ rather than covalent cadmium phosphorus sulfide. The consequence of the ionic metal–ligand interactions in the $CdPS_3$ layer is that these compounds exhibit an unique intercalation chemistry, not observed in the other layered compounds [20, 21]. The cation exchange intercalation reaction involves the insertion of a hydrated cationic species in the interlayer space with charge neutrality preserved by a loss of the divalent M^{+2} ions from the layer into solution. Mg–Al LDH solid is obtained by substituting Mg^{2+} ion in brucite $Mg(OH)_2$ by Al^{3+} ions. The layers of LDH solid are positively charged, and charge neutrality is maintained by the presence of anions in the interlamellar positions. This paper describes the self-assembly of exfoliated sheets of surfactant intercalated anionic and cationic clays to form a new layered solid with periodically alternating anionic and cationic layers. It is shown that a new crystalline solid with alternating $CdPS_3$—LDH layers can be obtained by mixing 1:1 dispersions of $CdPS_3$–DODMA with Mg–Al LDH-DDS followed by solvent evaporation. We have followed the reported procedure [22] that is intercalation of long-chain surfactant in the interlamellar space followed by sonication in nonpolar solvents, to achieve delamination of both $CdPS_3$ and Mg–Al LDH.

2 Results

The anionic clay $Mg_{0.66}Al_{0.33}(OH)_2\,(NO_3)_{0.33}$ (Mg–Al LDH) with $NO_3{}^-$ as the interlamellar anion was prepared by reported procedures [23]. Cadmium thiophosphate, $CdPS_3$, was prepared by reported procedures [24]. Intercalation of ionic surfactants in both Mg–Al LDH and $CdPS_3$ has been achieved by ion-exchange method [22]. When surfactant is intercalated, an increase in the interlayer separation has been observed depending on the chain length of the surfactant. The cationic surfactant intercalated in $CdPS_3$ was the di-octadecyldimethyl ammonium (DODMA) ion, while in the LDH, it was the dodecyl sulfate (DDS) anion. The XRD studies determined interlayer spacings of the intercalated $CdPS_3$–DODMA and Mg–Al LDH-DDS are 3.72 and 2.7 nm, respectively. This *d*-spacing suggests a bilayer arrangement of the intercalated surfactant chains in the interlamellar region [25, 26]. Delamination of the surfactant layered solids was achieved by dispersing the surfactant intercalated solid in non-polar organic solvent followed by sonication for 15 min.

The characterization of surfactant-intercalated solids, morphology, and size of the exfoliated nanosheets was examined by TEM, SAED, and tapping mode AFM. Bright field (BF) TEM images of the delaminated nanosheets are shown in Figs. 1 and 2. The TEM images show two-dimensional sheets of micron size width. The TEM images clearly show the sheets are having a well-defined morphology. For example, hexagonal morphology of Mg–Al LDH is clearly seen in Fig. 1. The tapping mode AFM image shows two dimensional sheets with an average thickness that corresponds to that of a single inorganic layer. For the layered double hydroxides, Mg–Al LDH, the sheets in the tapping mode AFM images are typically 0.6 nm. The thickness of the hydrotalcite sheet is 0.48 nm [27]. The selected area electron diffraction pattern of individual $CdPS_3$ sheets (shown in Fig. 2) exhibits the spotted diffraction patterns indicating their single crystal nature. The typical thickness of the $CdPS_3$–DODMA sheets in the AFM images are 0.62 nm. The thickness of a $CdPS_3$ sheet is 0.5 nm. The observation of single sheets in the AFM images indicates the complete exfoliation of solids, and the sheets are electrically neutral.

Mg–Al LDH-DDS and $CdPS_3$–DODMA were sonicated in chloroform separately for 15 min. The concentration of hydrotalcite or $CdPS_3$ in dispersion was 2 mg/mL. This dispersion were mixed and drop coated on a glass substrate and chloroform allowed to evaporate slowly. Samples with different molar ratios of Mg–Al LDH-DDS and $CdPS_3$–DODMA were prepared. X-ray diffraction patterns of the Mg–Al LDH-DDS, $CdPS_3$–DODMA are shown in the Fig. 3. It may be seen for the

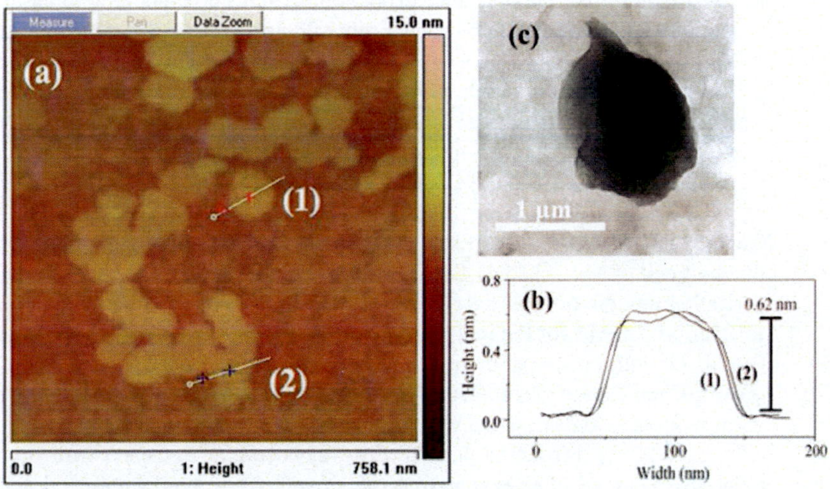

Fig. 1 **a** Tapping mode AFM images of the Mg–Al LDH-DDS nanosheets. **b** The height profiles of the sheets along the white lines marked on the images are shown. **c** Electron microscope images of Mg–Al LDH-DDS nanosheets. The inset shows the selected area diffraction pattern

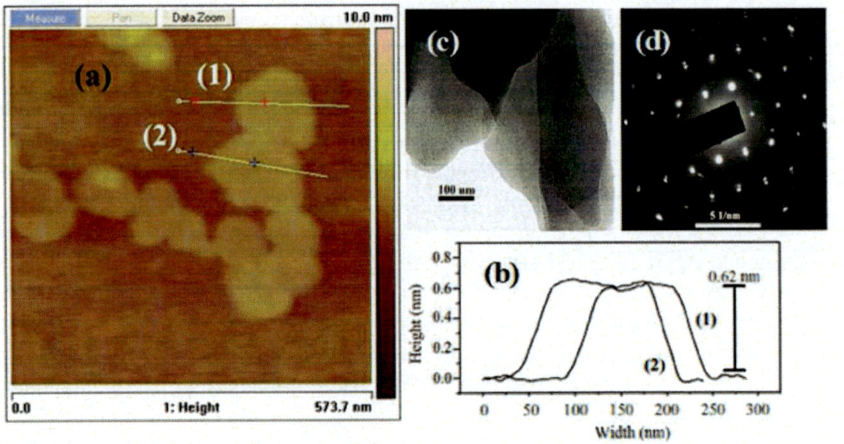

Fig. 2 **a** Tapping mode AFM images of the $CdPS_3$–DODMA nanosheets. **b** The height profiles of the sheets along the white lines marked on the images are shown. **c** Electron microscope images of $CdPS_3$–DODMA nanosheets. **d** The selected area diffraction pattern of $CdPS_3$–DODMA nanosheets

mixture of $CdPS_3$–DODMA and Mg–Al LDH-DDS dispersion a new set of reflections (shaded peaks in Fig. 3a) are observed. At a molar ratio of 1:1, peaks corresponding to neither the surfactant intercalated Mg–Al LDH or $CdPS_3$ are observed and only the new set of reflections are observed. For this composition, the peaks in XRD pattern may be indexed as (00*l*) reflections corresponding to an interlayer spacing of 5.48 nm. Figure 3b shows a plot of d-spacings associated with each of the (00*l*) reflections against 1/*l* for the 1:1 molar ratio. The plot of the d-spacing versus 1/*l* is linear confirming single phase with a structure that is periodically ordered. The composition of the metal ions in the hybrid that gave a unique XRD pattern was established by ICP-OES analysis. Water content and surfactant concentrations were obtained from TGA and elemental analysis, respectively. The unit cell formula of hybrid is $Mg_{0.37}Al_{0.18}Cd_{0.44}P_{0.56}S_{1.68}(OH)_{1.11}(DDS)_{0.18}(DODMA)_{0.12}\cdot 0.92H_2O$. This formula is close to the formula calculated by considering the combination of one unit cell of Mg–Al LDH-DDS and one $CdPS_3$–DODMA.

Reflex module of Materials Studio package [28] was used for calculating X-ray diffraction patterns for the inorganic layered lattice. The Mg–Al LDH structure was calculated with lattice parameters a = b = 0.304 nm, c = 2.281 nm, $\alpha = \beta = 90°$, $\gamma = 120°$, and space group R3m (ICSD No. 81963) [29]. The interlayer distance of 2.7 nm for Mg–Al LDH-DDS was achieved by dilating 'c' parameter to 8.1 nm. The calculated X-ray diffraction pattern (Fig. 4b) is similar to the experimental pattern (Fig. 4a) recorded after evaporating the solvent from the delaminated dispersion. The $CdPS_3$ structure was generated with lattice parameters a = 0.621 nm, b = 1.763 nm, c = 0.686 nm, $\alpha = \gamma = 90°$, $\beta = 107.58°$, and space group C12/m1 (ICSD No. 61393) [30]. Some of the cadmium ions were removed from the layer,

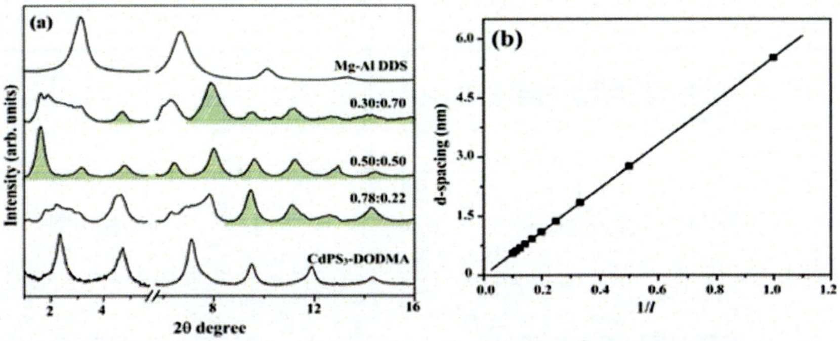

Fig. 3 **a** X-ray diffraction patterns of the solid obtained on solvent evaporation from different molar ratio mixtures of dispersions of Mg–Al LDH-DDS and $CdPS_3$–DODMA in chloroform. **b** The linear variation of the d-spacing versus 1/*l* for the 1:1 hybrid

and the 'c' parameter dilated to 3.72 nm. The calculated X-ray diffraction pattern (Fig. 5b) is similar to the experimental pattern (Fig. 5a) recorded after evaporating the solvent from the delaminated dispersion of $CdPS_3$–DODMA. The hybrid structure was generated by placing a Mg–Al LDH layer midway between two $CdPS_3$ layers separated by 5.48 nm. This would correspond to a hybrid of $CdPS_3$ and hydrotalcite layers. The calculated X-ray diffraction pattern for a structure with Mg–Al LDH and $CdPS_3$ sheets alternate in a periodic fashion with an interlayer repeat of 5.48 nm is similar to the experimentally recorded diffraction pattern of the 1:1 hybrid (Fig. 6). If the anchored surfactant forms bilayers like in their parent Mg–Al LDH-DDS and $CdPS_3$–DODMA compounds, an interlayer spacing of 6.36 nm may be expected for the 1:1 hybrid. The observed spacing of 5.48 nm suggests that the surfactant bilayers adopt a interdigitated bilayer arrangement.

It has been already shown that perfectly periodic alternating layered structure can be prepared by mixing the surfactant-tethered exfoliated nanosheets of hydrotalcite and montmorillonite in a nonpolar solvent followed by evaporation of the solvent. In previous report, the existence of the electrostatic interaction between the 'tails' of the oppositely charged surfactant chains anchored to the hydrotalcite and montmorillonite sheets has been established [31]. Similar analogy can be applied here in the formation of hydrotalcite–$CdPS_3$ hybrid. The opposite residual charges present on the 'tails' of the DDS and DODMA alkyl chains driving the hybrid formation between Mg–Al LDH and $CdPS_3$ sheets.

3 Conclusion

In summary, it has been shown that when dispersions of the surfactant anchored MgAl LDH and $CdPS_3$ solids are mixed and the solvent allowed to evaporate, a new solid is obtained in which the anionic and cationic sheets alternate in a periodic fashion with

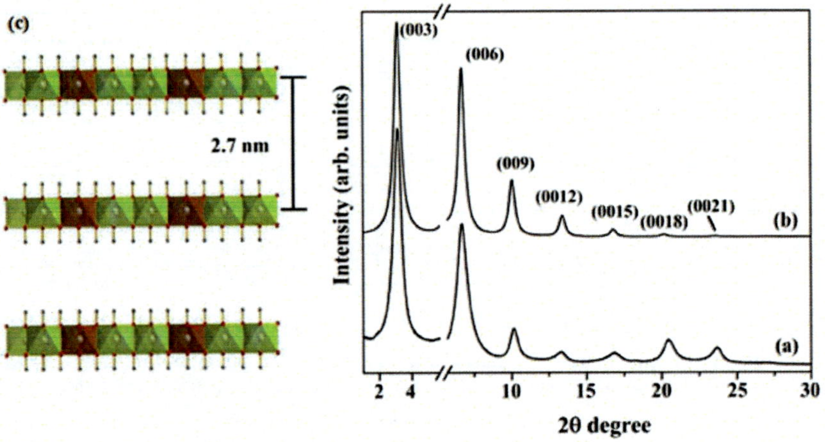

Fig. 4 Comparison of the experimental and calculated diffraction patterns of Mg–Al LDH-DDS. **a** The experimental diffraction pattern and **b** calculated pattern of Mg–Al LDH sheets with an interlayer spacing of 2.7 nm as shown in the panel **c** on the left

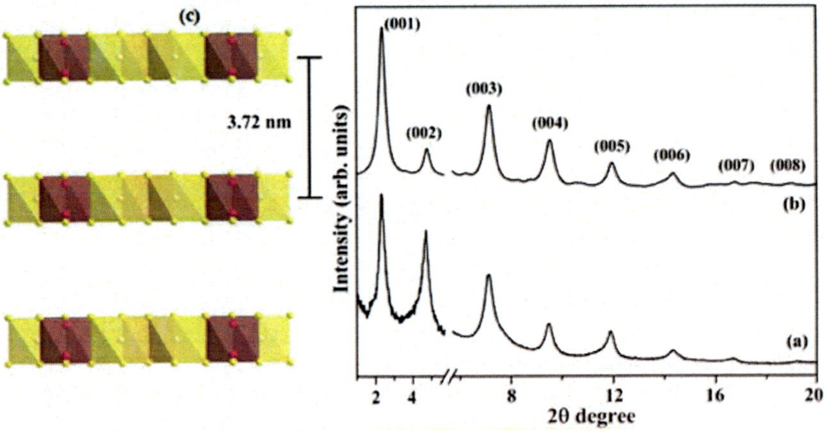

Fig. 5 Comparison of the experimental and calculated diffraction patterns of $CdPS_3$–DODMA. **a** The experimental diffraction pattern and **b** calculated pattern of $CdPS_3$–DODMA sheets with an interlayer spacing of 3.72 nm as shown in the panel **c** on the left

a repeat length different from that of the starting solids. In this method, electrostatic attractive forces between the tails of the anion and cationic surfactant chains anchored to the cationic and anionic inorganic sheets ensures self-assembly of a perfectly periodically alternating cationic–anionic layered structure. The procedure presented here may easily be extended to other layered solids like graphene for creating new superstructures from 2D exfoliated inorganic nanosheets by self-assembly.

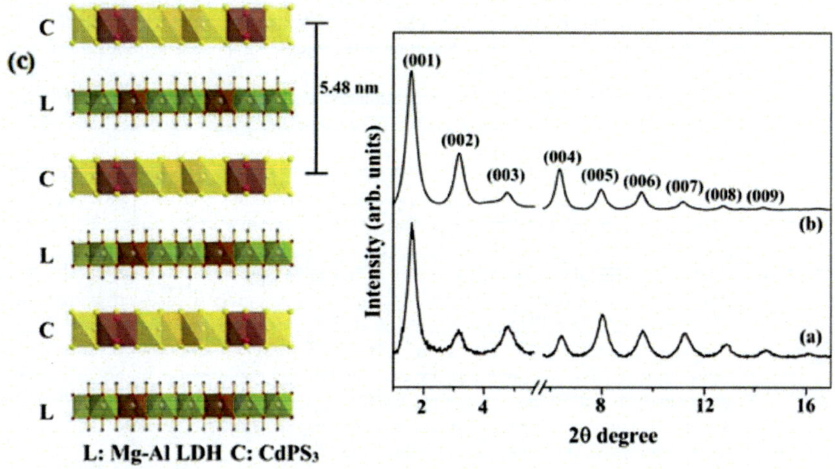

Fig. 6 Comparison of the experimental and calculated diffraction patterns of Mg–Al LDH-DDS—$CdPS_3$–DODMA hybrid. **a** The experimental diffraction pattern and **b** calculated pattern considering a periodically ordered alternating hydrotalcite–$CdPS_3$ stack with a repeat length of 5.48 nm as shown in the panel **c** on the left

References

1. Desiraju GR, Vittal JJ, Ramanan A (2011) Crystal engineering. A text book. World Scientific Publishing, Singapore
2. Coleman JN et al (2011) Two-dimensional nanosheets produced by liquid exfoliation of layered materials. Science 331(6017):568–571
3. Golberg D (2011) Nanomaterials: exfoliating the inorganics. Nature Nanotech 6:200–201
4. Norrish K (1954) The swelling of montmorillonite. Discuss Faraday Soc 18:120–134
5. Wang Q, O'Hare D (2012) Recent advances in the synthesis and application of layered double hydroxide (LDH) nanosheets. Chem Rev 112(7):4124–4155
6. Novoselov KS et al (2005) Two-dimensional atomic crystals. Proc Natl Acad Sci 102(30):10451–10453
7. Pinnavaia TJ, Beall GW (2000) Polymer-clay nanocomposites. Wiley, New York
8. Vaia RA, Giannelis EP (2001) Polymer nanocomposites: status and opportunities. MRS Bull 26(5):394–401
9. Smith RJ et al (2011) Large-scale exfoliation of inorganic layered compounds in aqueous surfactant solutions. Adv Mater 23(34):3944–3948
10. Ma R, Sasaki T (2010) Nanosheets of oxides and hydroxides: ultimate 2D charge bearing functional crystallites. Adv Mater 22(45):5082–5104
11. Coronado E et al (2010) Coexistence of superconductivity and magnetism by chemical design. Nature Chem 2:1031–1036
12. Srivastava S, Kotov NA (2008) Composite layer-by-layer (LBL) assembly with inorganic nanoparticles and nanowires. Acc Chem Res 41(12):1831–1841
13. Ma R, Sasaki T (2012) Synthesis of LDH nanosheets and their layer-by-layer assembly. Recent Pat Nanotech 6(3):159–168
14. Li L, Ma R, Ebina Y, Fukuda K, Takada K, Sasaki T (2007) Layer-by-layer assembly and spontaneous flocculation of oppositely charged oxide and hydroxide nanosheets into inorganic sandwich layered materials. J Am Chem Soc 129(25):8000–8007

15. Lee T, Hoon T, Minsu Gu M, Jung YK, Lee W, Uk Lee J, Seong DG, Kim B-S (2015) Layer-by-layer assembly for graphene-based multilayer nanocomposites: synthesis and applications. Chem Mater 27(11):3785–3796
16. Li B-W, Osada M, Ozawa TC, Ebina Y, Akatsuka K, Ma R, Funakubo H, Sasaki T (2010) Engineered interfaces of artificial perovskite oxide superlattices via nanosheet deposition process. ACS Nano 4(11):6673–6680
17. Amratisha K, Ponchai J, Kaewurai P, Pansa-ngat P, Pinsuwan K, Kumnorkaew P, Ruankham P, Kanjanaboos P (2020) Layer-by-layer spray coating of a stacked perovskite absorber for perovskite solar cells with better performance and stability under a humid environment. Opt Mater Express 10(7):1497–1508
18. Chalmpes N, Kouloumpis A, Zygouri P, Karouta N, Spyrou K, Stathi P, Tsoufis T, Georgakilas V, Gournis D, Rudolf P (2019) Layer-by-layer assembly of clay-carbon nanotube hybrid superstructures. ACS Omega 4(19):18100–18107
19. Gunjakar JL, Kim TW, Kim HN, Kim IY, Hwang SJ (2011) Mesoporous layer-by-layer ordered nanohybrids of layered double hydroxide and layered metal oxide: highly active visible light photocatalysts with improved chemical stability. J Am Chem Soc 133(38):14998–15007
20. Joy PA, Vasudevan S (1992) The intercalation reaction of pyridine with manganese thiophosphate, $MnPS_3$. J Am Chem Soc 114(20):7792–7801
21. Clement R (1981) Intercalation of potentially reactive transition-metal complexes in the lamellar manganese phosphide sulfide host lattice. J Am Chem Soc 103(23):6998–7000
22. Naik VV, Ramesh TN, Vasudevan S (2011) Neutral nanosheets that gel: exfoliated layered double hydroxides in toluene. J Phys Chem Lett 2(10):1193–1198
23. Olanrewaju J, Newalkar BL, Mancino C, Komarneni S (2000) Simplified synthesis of nitrate form of layered double hydroxide. Mater Lett 45(6):307–310
24. Klingen W, Ott R, Hahn HZ (1973) Uber die Darstellung und Eigenschaften von Hexathio- und Hexaselenohypodiphosphaten. Anorg Allg Chem 396(3):271–273
25. Naik VV, Chalasani R, Vasudevan S (2011) Composition driven monolayer to bilayer transformation in a surfactant intercalated Mg–Al layered double hydroxide. Langmuir 27(6):2308–2316
26. Venkataraman NV, Vasudevan S (2002) Characterization of alkyl chain conformation in an intercalated cationic lipid bilayer by IR spectroscopy. J Phys Chem B 106(32):7766–7773
27. Meyn M, Beneke K, Lagaly G (1990) Anion-exchange reactions of layered double hydroxides. Inorg Chem 29(26):5201–5207
28. Materials Studio (2010) Version 5.5.0.0. Accelrys Software Inc
29. Bellotto M, Rebours B, Clause O, Lynch J, Bazin D, Elkaïm E (1996) A reexamination of hydrotalcite crystal chemistry. J Phys Chem 100(20):8527–8534
30. Ouvrard G, Brec R, Rouxel J (1985) Structural determination of some MPS_3 layered phases (M = Mn, Fe Co, Ni and Cd). Mat Res Bull 20(10):1181–1189
31. Chalasani R, Gupta A, Vasudevan S (2013) Engineering new layered solids from exfoliated inorganics: a periodically alternating hydrotalcite–montmorillonite layered hybrid. Sci Rep 3:1–8, 3498

Deposition Time-Dependant Properties of PbS Thin Films

Srinivasa Reddy Tippasani, S. Vijaya Krishna, and M. C. Santhosh Kumar

Abstract PbS thin films are deposited on soda-lime glass substrates using simple and inexpensive chemical bath deposition (CBD) method. We investigated the structural, morphological, and optical properties of PbS thin films as a function of deposition time. XRD and Raman analysis confirm the phase of PbS thin films. The scanning electron microscopy (SEM) examined to study the morphological properties of the films. The optical properties such as transmission, absorption, and band gap of the films were investigated by UV-Vis-NIR spectrometer. The optical direct band gap values (1.10–0.93 eV) are estimated using the absorption data.

Keywords PbS thin films · CBD method · Structural and optical characterizations · Band gap

1 Introduction

In the recent years, deposition of metal chalcogenide materials in particular binary (ZnS, SnS, PbS) and ternary (Cu_xSnS_y, AgInS) has attracted great interest in the optoelectronic applications due to their considerable optical and electrical properties. Chalcogenide thin films have great consideration in the deposition of large area with variety of surfaces. Lead sulfide (PbS) is a well-known IV–VI group of binary semiconductor with a p-type direct energy band gap. It can be varied from 0.41 to 2.3 eV with large excitation Bohr radius of 18 nm [1, 2]. It can be used for solar coatings [3], infrared detector [4], gas sensors [2], photoresistors, and solar absorbers [5]. Moreover, the optical constants of PbS semiconductor like refractive index and extinction coefficient are suitable for the optical sensor devices due to their narrow energy band gap [0.37–0.4 eV at 300 K] [6]. In general, PbS thin films can be prepared

S. R. Tippasani (✉)
Department of Physics, P.B. Siddhartha College of Arts and Science, Vijayawada, Andhra Pradesh 520010, India
e-mail: tsrinivasareddy@pbsiddhartha.ac.in

S. V. Krishna · M. C. S. Kumar
Optoelectronic Materials and Devices Laboratory, Department of Physics, National Institute of Technology, Tiruchirappalli, Tamil Nadu 620015, India

N. M. Rao et al. (eds.), *Advanced Nanomaterials and Their Applications*,
Springer Proceedings in Materials 22, https://doi.org/10.1007/978-981-99-1616-0_6

by chemical and physical methods such as Vacuum evaporation [7], RF sputtering [8], pulsed laser deposition [9], Successive ionic layer adsorption and reaction (SILAR) [4, 10], chemical bath deposition (CBD) [11, 12], hydrothermal method [13] and electrodeposition [14].

In this work, chemical bath deposition method was employed to deposit PbS thin films at various deposition times from 30 to 120 min. The deposition time-dependent study on the physical properties of PbS thin films reported here.

2 Experimental

Lead sulfide (PbS) thin films were deposited using simple and inexpensive chemical bath deposition technique (CBD). Figure 1 shows the schematic diagram of CBD method. The glass substrates cleaned with running water and immersed in diluted HCl solution for 24 h. Further, these glass substrates were cleaned with detergent solution and thoroughly cleaned with double distilled water and fallowed by ultrasonic process for 30 min. These slides were further cleaned by acetone and dried with air dryer, prior to the deposition. Lead acetate [$Pb(CH_3COO)_2$], thiourea [$SC(NH_2)_2$], ammonia solution NH_4OH, and triethanolamine (TEA) were used as precursor solution for the PbS thin films. The chemical bath is prepared by 50 ml of 0.1 M lead acetate as the lead source, 50 ml of 0.2 M thiourea as the sulfur source, and 5 ml of ammonia solution and few drops of TEA. Triethanolamine was used as the complex agent. The temperature of the reactive solution was maintained at 60 °C, and pH was adjusted at 12 by adding ammonia solution. The PbS thin films are deposited on glass substrates at different deposition times from 30 to 120 min at an interval of 30 min. In CBD method, the formation of PbS depends on the ion-by-ion mechanism. In this method, the growth of PbS depends on the concentration of lead and sulfur ions. However, PbS films are possible only when the ionic product [Pb^{+2}] [S^{2-}] is always higher than the solubility product of PbS [15]. The possible chemical reaction for the formation of PbS thin films on the glass substrate is as follows [16]

$$Pb(CH_3COO)_2 \cdot 2H_2O + TEA \leftrightarrow [Pb(TEA)]^{2+} + 2(CH_3OO)^-$$

$$[Pb(TEA)]^{2+} \leftrightarrow Pb^{2+} + TEA$$

The metal complex $[Pb(TEA)]^{2+}$ is formed due the addition of complex agent TEA. This metal complex agent leads to avoid the formation of precipitation state $Pb(OH_2)$ in the chemical bath and also control the reaction rate. The decomposition of the thiourea in aqueous solution described by the following reaction mechanism [16, 17],

$$SC(NH_2)_2 + OH^- \rightarrow CH_2N_2 + H_2O + HS^-$$

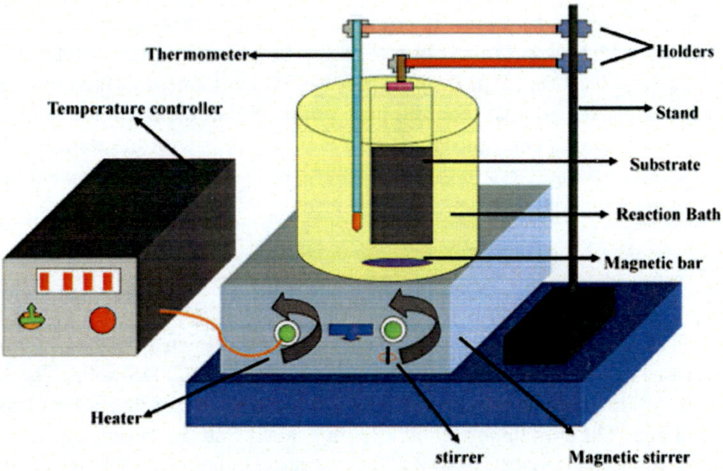

Fig. 1 Schematic diagram of CBD method

$$HS^- + OH^- \leftrightarrow S^{2-} + H^2O$$

$$Pb^{2+} + S^{2-} \rightarrow PbS$$

The resultant films are very adherent and smooth on the substrate with gray color surface like mirror. The color of the deposited thin films changes from light gray to dark gray with increase of deposition time from 30 to 120 min.

X-ray diffraction was employed to study the structural features of all the films. Raman spectroscopy was used to confirm the phases of the films. The morphological features of the films were observed by ultra 555 scanning electron microscopy (SEM). The chemical composition of the films was characterized by the energy dispersive spectroscopy (EDS). JASCO V-670 UV-ViS-NIR spectrometer was used to record the optical measurements of the films.

3 Results and Discussion

3.1 Structural Studies

Figure 2 shows the XRD pattern of the deposited films at different deposition times from 30 to 120 min. It is clear that all the films show the polycrystalline in nature with orientation of planes (111), (200), (220), and (311). The preferential orientation along (200) plane indicates that the good crystalline nature with lower crystalline defects.

The intensity of the preferred orientation (200) increases with increase of deposition time. Thoidi et al. [17] also observed a similar variation in (200) plane with deposition time. It is clearly exhibits the pure phase of PbS without any other metallic impurities. Rajathi et al. [18] also reported pure phase of PbS using chemical bath deposition. The diffraction peaks and planes are well matches with the standard JCPDS-5592. The lattice parameters of the cubic phase of PbS thin films are calculated using the relation.

$$\frac{1}{d^2} = \frac{h^2 + k^2 + l^2}{a^2} \tag{1}$$

The estimated lattice parameter values are shown in Table 1, which are well matches with the reported values [15] and standard JCPDS-5592. The lattice parameter "a" increases from 5.81 to 5.91 Å with increase of deposition time from 30 to 120 min. The film deposited at 120 min shows that the lattice parameter is closer to the bulk PbS crystals (5.930 Å). The variation in lattice parameter (a) was due to the presence of strain and dislocation density of the films. The average crystallite size (D), microstrain (ε), and dislocation density (δ) of the deposited films are calculated by following equations.

$$D = \frac{0.94\lambda}{\beta \cos\theta} \tag{2}$$

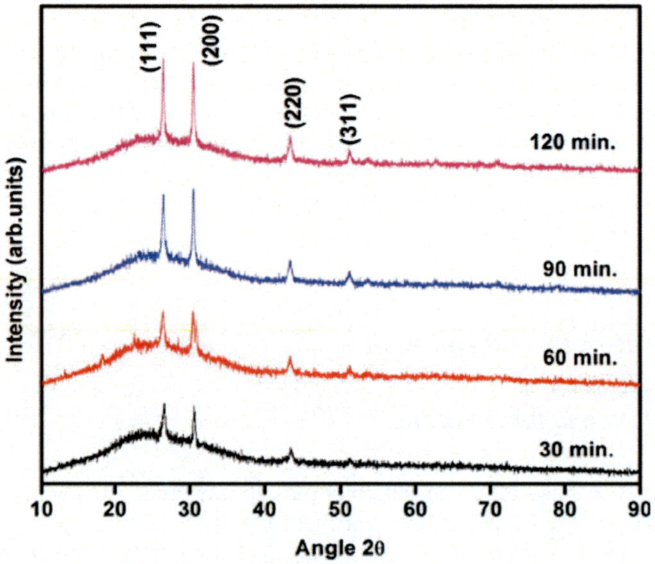

Fig. 2 X-ray diffraction pattern of the PbS films

Table 1 Structural and lattice parameters of the deposited films

Deposition time (min)	FWHM (β) (radians)	Average crystallite size (D) nm	Microstrain (ε)	Dislocation density (δ) × 10^{16} (m^{-2})	(hkl)	Lattice parameters (Å) ($a = b = c$)
30	0.3531	24	0.067	7.20	(111)	5.81
60	0.3217	27	0.061	5.792	(200)	5.85
90	0.2735	31	0.052	4.271	(220)	5.89
120	0.2537	33	0.048	3.691	(311)	5.91

$$\varepsilon = \frac{\beta \cos \theta}{4} \quad (3)$$

$$\delta = \frac{15\varepsilon}{aD} \quad (4)$$

The estimated values are shown in Table 1. The average crystallite size (D) of the deposited films increases from 24 to 33 nm with increase of deposition times from 30 to 120 min. Similar results are reported in other studies [19–21]. The microstrain (ε) and dislocation density (δ) values of the deposited films are found to decrease with increase of deposition time which strongly favors that increment in average crystallite size (D) of the deposited films. The sample (120 min) shows lower values of microstrain (ε) and dislocation density (δ) which might be due to the larger nucleation growth rate compared to other samples.

3.2 Raman Studies

Figure 3 shows the Raman modes of the PbS thin films recorded at room temperature in the range of 80–1100 cm^{-1} using 514 nm laser source. From the figure, noticed that the Raman bands are recorded at 92, 140, and 967 cm^{-1}. The Raman mode at 92 cm^{-1} is a characteristic peak of the PbS thin films, which is attributed to the combination of lattice mode vibrations (longitudinal (LA) + transverse acoustic phonon (TA) modes). Gode et al. [22] also reported a similar peak position using 633 nm laser line for PbS thin films deposited by SILAR method. The strong Raman band observed at 140 cm^{-1} for the samples above 60 min. This Raman peak was attributed due to the scattering by transverse acoustic and optical (TA + TO) phonon modes. It is comparable to the reported values in PbS crystals [17, 23–25]. The low intensity peak at 967 cm^{-1} may be originated from the sulfates group ($PbSo_4$), which is formed the due to laser-induced degradation. A similar Raman peak was reported in PbS thin films [12, 26].

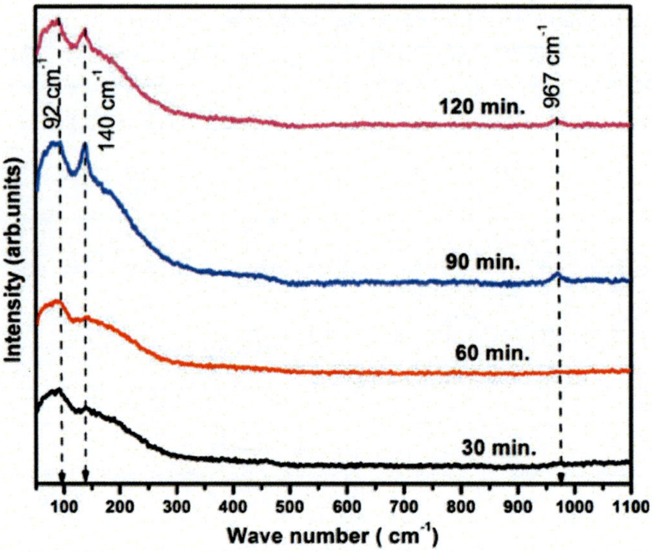

Fig. 3 Raman spectra of the PbS thin films

3.3 Morphological Studies

Figure 4a–d depicts the FESEM images of the deposited PbS thin films. It is clear that these micrographs exhibit homogeneous crystallites which are uniformly covered across the surface of the film without any cracks. The variation in the average grain size of the films can be understood by deposition times. The high resolution images of the samples are shown in the insert of Fig. 4a–d. It is observed that the most of the crystallites are in cubic shape. It is noticed that the average grain size of the films was increased from 179 to 225 nm with increase of deposition time 30–120 min. These values are well matches with the values reported by Tohidi et al. [17]. EDS analysis was carried out to confirm the existence of elements in the samples. Figure 4e shows the EDS spectrum of deposition time 120 min sample. It is clear that the film contains peaks corresponding lead (Pb) and sulfur (S). The observed atomic percentage of the sample was 44.96 and 55.04 at% corresponding Pb and S. Ambios XP surface profilometer was used to estimate the thickness (90–250 nm) of the PbS films.

3.4 Optical Studies

Figure 5a depicts the optical transmittance spectra of the deposited samples at room temperature. The deposited films exhibit decrease in transparency from 45 to 5% below 1200 nm and above that the transmittance was systematically increases with

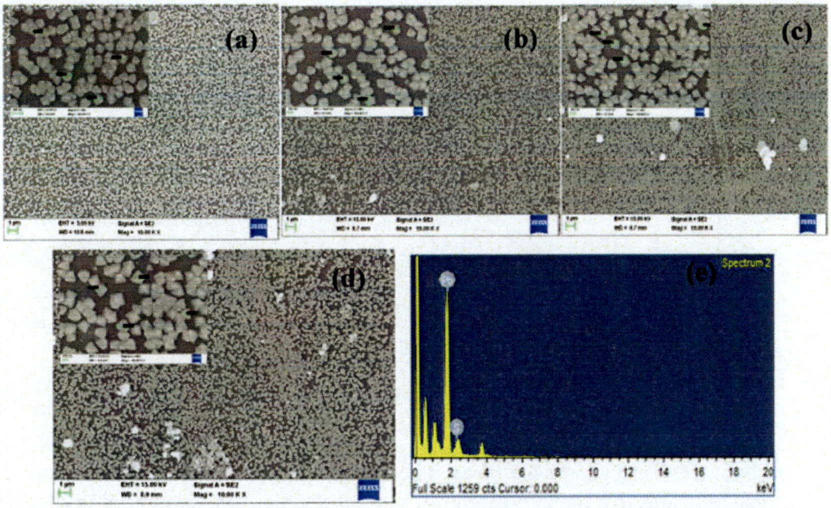

Fig. 4 SEM micrographs of PbS thin films **a** 30 min **b** 60 min **c** 90 min **d** 120 min **e** EDS spectrum of 120 min sample. (Inset: high resolution images of the deposited films)

increase of wave length. These values are lower than the reported transparency values of chemically deposited PbS thin films (60–40%) by Tohidi et al. [17]. The lower transmittance and higher absorption values in the visible region suggest that these deposited films are more suitable for optoelectronic applications. The variation in the transmittance of the deposited films can be associated to the thickness of the deposited films. In this study, the thickness of the deposited films increases with increase of deposition time. The deposited films show decrease in transparency in the NIR region with increase of deposition time. The film deposited at 120 min exhibits lower transmittance in the NIR region, which might be due to the higher thickness of the films. Figure 5b shows the absorption spectra of the deposited films. From the figure, it is noticed that all the deposited films exhibit steep absorption edge, which is shifted toward the higher wave length with increase of deposition time from 30 to 120 min. The variation in absorption edge with wave length indicates that the reduction in energy band gap. A similar shift in absorption edge was observed for other metal chalcogenide materials [27, 28].

The optical direct band gap of the films calculated from the absorption data.

$$\alpha h\upsilon = M\left(h\upsilon - E_g\right)^n$$

where α is absorption coefficient, M is a constant, E_g is the optical band gap, and n denotes the nature of the transition ($n = 1/2$, $3/2$ for allowed direct band gap and $n = 2$, 3 are for indirect band gap). Figure 6 shows the Tauc's plot in order to estimate the optical direct band gap of the deposited films. The energy band gap of the deposited films estimated as the intercept of tangential linear portion on the horizontal axis $h\upsilon$

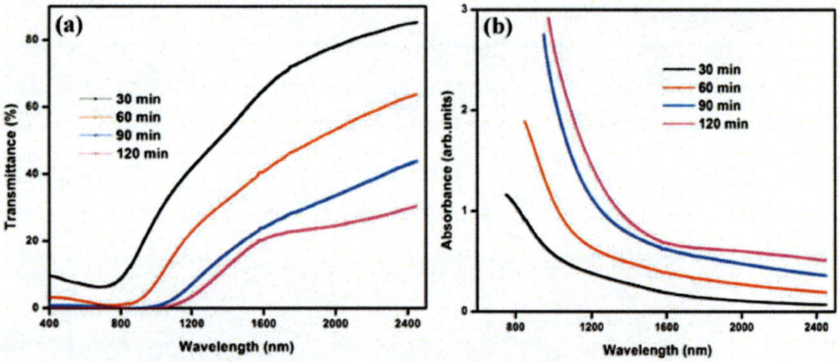

Fig. 5 **a** Transmittance spectra and **b** absorption spectra of the PbS thin films

above the band gap of the films. The straight line behavior with $h\upsilon$ above the band gap indicates that the deposited films are direct band gap nature. A similar band gap nature of the PbS thin films reported earlier [2, 16, 17]. The estimated energy band gap values are decreased from 1.10 to 0.93 eV with increase of deposition time. These values are better than the reported energy band gap values of PbS thin films using chemical bath deposition [2, 17–19, 29]. Hone et al. [16] also reported a similar trend in the band gap of the films by chemical bath deposition method.

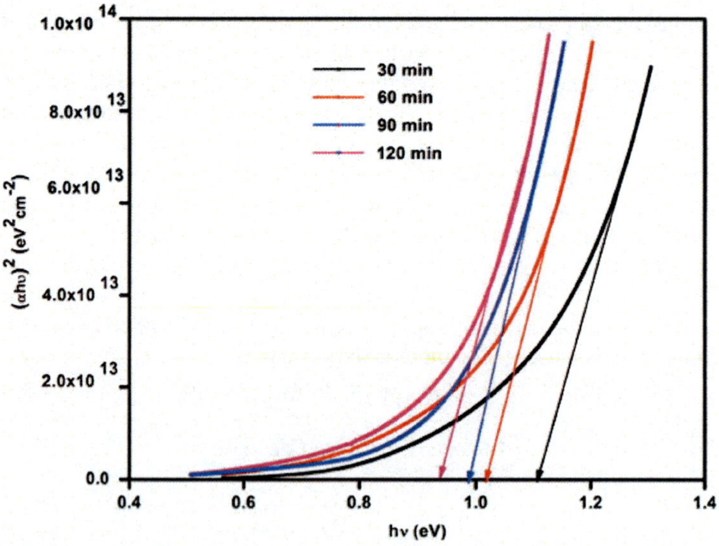

Fig. 6 Tauc's plots of PbS thin films

The variation in optical band of the films may be depend on average crystallite size, microstrain, composition, and defects. The estimated band gap values are a significant blue shift from the bulk value 0.41 eV at room temperature. In this case, the decrease in band gap of the films with deposition time might be due to the quantum size effect exhibited by nanoparticles size dispersion in the films [6, 16, 30]. From the optical band gap analysis, it is clear that the deposited PbS thin films are prospective absorber material for solar cell applications.

4 Conclusions

PbS thin films have been prepared by chemical bath deposition. XRD and Raman analysis confirm the pure PbS phase without any other phases. The average crystallite size of the deposited films was increased from 24 to 33 nm with increase of deposition times from 30 to 120 min. From SEM analysis, it is clear that the cubical shapes of the crystallites are distributed with homogeneously on the surface of the film. The presence of lead (Pb) and sulfur (S) elements in the deposited films is identified by EDS analysis. The shift in absorption edge with deposition time was studied using UV-Vis-NIR spectrometer. The direct band gap of the films was decreases from 1.10 to 0.93 eV with increase of deposition time from 30 to 120 min. From optical analysis, these films can be used to achieving higher efficiency solar cells as an absorber layer.

References

1. Ji J, Ji H, Wang J, Zheng X, Lai J, Liu W, Li T, Ma Y, Li H, Zhao S, Jin Z (2015) Deposition and characteristics of PbS thin films by an in-situ solution chemical reaction process. Thin Solid Films 590:124–133
2. Yucel E, Yucel Y, Beleli B (2015) Optimization of synthesis conditions of PbS thin films grown by chemical bath deposition using response surface methodology. J Alloy Compd 642:63–69
3. Touati B, Gassoumi A, Alfaify S, Turki NK (2015) Optical, morphological and electrical studies of Zn:PbS thin films. Mater Sci Semicond Process 34:82–87.
4. Güneria E, Gode F, Cevik S (2015) Influence of grain size on structural and optic properties of PbS thin films produced by SILAR method. Thin Solid Films 589:578–583
5. Gunes S, Fritz KP, Neugebauer H, Sariciftci NS, Kumar S, Scholes GD (2007) Hybrid solar cells using PbS nanoparticles. Sol Energy Mater Sol Cells 91:420–423
6. Sadovnikov SI, Gusev AI (2013) Structure and properties of PbS films. J Alloy Compd 573:65–75
7. Kumar S, Sharma TP, Zulfequar M, Husainb M (2003) Characterization of vacuum evaporated PbS thin films. Phys B 325:8–16
8. Filho JCDS, Marques FC (2019) Structural and optical temperature-dependent properties of PbS thin films deposited by radio frequency sputtering. Mater Sci Semicond Process 91:188–193
9. Atwa DMM, Azzouz IM, Badr Y (2011) Optical, structural and optoelectronic properties of pulsed laser deposition PbS thin film. Appl Phys B 103:161–164

10. Yucel E, Yucel Y, Beleli B (2015) Process optimization of deposition conditions of PbS thin films grown by a successive ionic layer adsorption and reaction (SILAR) method using response surface methodology. J Cryst Growth 422:1–7
11. Abdallah B, Hussein R, Kafri NA, Zetoun W (2019) PbS thin films prepared by chemical bath deposition: effects of concentration on the morphology, structure and optical properties. Iran J Sci Technol, Trans A 43:1371–1380
12. Bhatt SV, Deshpande MP, Soni BH, Garg N, Chaki S (2014) Chemical bath deposition of lead sulphide (PbS) thin film and their characterization. Solid State Phenom 209:111–115
13. Niasari MS, Ghanbari D, Estarki MRL (2012) Star-shaped PbS nanocrystals prepared by hydrothermal process in the presence of thioglycolic acid. Polyhedron 35:149–153
14. Mathewsa NR, Chavez CA, Jacome MAC, Antonio JAT (2013) Physical properties of pulse electrodeposited lead sulfide thin films. Electrochim Acta 99:76–84
15. Gode F, Guneri E, Emen FM, Kafadar VE, Unlu S (2014) Synthesis, structural, optical, electrical and thermoluminescence properties of chemically deposited PbS thin films. J Lumin 147:41–48
16. Hone FG, Dejene FB (2018) Six complexing agents and their effects on optical, structural, morphological and photoluminescence properties of lead sulphide thin films prepared by chemical route. J Lumin 201:321–328
17. Tohidi T, Ghaleh KJ, Namdar A, Ghaleh R (2014) Comparative studies on the structural, morphological, optical, and electrical properties of nanocrystalline PbS thin films grown by chemical bath deposition using two different bath compositions. Mater Sci Semicond Process 25:197–206
18. Rajathi S, Kirubavathi K, Selvaraju K (2017) Structural, morphological, optical, and photoluminescence properties of nanocrystalline PbS thin films grown by chemical bath deposition. Arab J Chem 10:1167–1174
19. Abdallah B, Ismail A, Kashoua H, Zetoun W (2018) Effects of deposition time on the morphology, structure, and optical properties of PbS thin films prepared by chemical bath deposition. J Nanomater 2018:1–8
20. Rex Rosario S., Kulandaisamy I, Kumar KDA, Arulanantham AMS, Valanarasu S, Youssef MA, Awwad NS (2019) Deposition of p-type Al doped PbS thin films for heterostructure solar cell device using feasible nebulizer spray pyrolysis technique. Phys B 575:411704
21. Liu M, Zhan Q, Li W, Li R, He Q, Wang Y (2019) Effect of Zn doping concentration on optical band gap of PbS thin films. J Alloy Compd 792:1000–1007
22. Gode F, Baglayan O, Guneri E (2015) p-Type nanostructure PbS thin films prepared by the SILAR method. Chalcogenide Lett 12:519–528
23. Sukarova BM, Najdoski M, Grozdanov I, Chunnilall CJ (1997) Raman spectra of thin solid films of some metal sulfides. J Mol Struct 410–411:267–270
24. Rajashree C, Balu AR (2016) Tuning the physical properties of PbS thin films towards optoelectronic applications through Ni doping. Optik 127:8892–8898
25. Smith GD, Firth S, Clark RJH (2002) First- and second-order Raman spectra of galena (PbS). J Appl Phys 92:4375
26. Vankhade D, Chaudhuri TK (2019) Effect of thickness on structural and optical properties of spin-coated nanocrystalline PbS thin films. Opt Mater 98:109491
27. Reddy TS, Santhosh Kumar MC (2016) Effect of substrate temperature on the physical properties of co-evaporated Sn_2S_3 thin films. Ceram Int 42:12262
28. Reddy TS, Santhosh Kumar MC (2021) Temperature-dependent properties of Co-evaporated CuS thin films. Braz J Phys 51:1575–1583
29. Tohidi T, Ghaleh KJ (2015) Effect of TEA on photoluminescence properties of PbS nanocrystalline thin films. Appl Phys A 118:1247–1258
30. Fan L, Wang P, Guo Q, Han H, Li M, Chen Z, Zhao H, Zhang D, Zheng Z, Yang J (2015) Ultrasound-modulated microstructure of PbS film in ammonia-free chemical bath deposition. RSC Adv 5:10018

Investigation on Surface Trap Characteristics of Water-Diffused Al–Epoxy Nanocomposites

Chillu Naresh and Ramanujam Sarathi

Abstract During the operation of an HVDC cable in an offshore long-distance power transmission mode, the insulation on the cable's accessories could age and degrade if they absorbed moisture from the surrounding air, such as the diffusion caused by rain or fog condensation. Surface characteristics of water-diffused insulating materials should be studied since the degradation commences there. Surface potential measurements are employed to characterise the surface of water-diffused Al–epoxy nanocomposites, and these nanocomposites are compared to their unaged counterparts to provide a reference for surface characterisation. The diffusion coefficients of aged nanocomposites were found to decrease experimentally with the controlled addition of nanofiller. In spite of the fact that water diffusion nanocomposites have substantially reduced the surface potential characteristics, the incorporation of nanofillers has been shown to slightly improve these characteristics.

Keywords Aluminium · Diffusion coefficient · Epoxy · HVDC · Mean lifetime · Nanocomposites · Surface charge · Trap depth · Water diffusion

1 Introduction

Transmission of large amounts of electricity across greater distances is made possible by high voltage direct current (HVDC) power systems. HVDC lines have lower costs and less losses for long-distance power transmission than AC transmission. Long-distance power transmission and distribution via HVDC systems have risen to prominence in recent years as a result of efficient connectivity between asynchronous AC networks, increased integration of renewable energy, satisfaction of power trading between countries, and the desire to transmit power over longer distances. It allows

C. Naresh (✉)
Department of Electronics Engineering, VIT-AP University, Amaravati 522237, India
e-mail: naresh.ch@vitap.ac.in

R. Sarathi
Department of Electrical Engineering, Indian Institute of Technology-Madras, Chennai 600036, India

N. M. Rao et al. (eds.), *Advanced Nanomaterials and Their Applications*,
Springer Proceedings in Materials 22, https://doi.org/10.1007/978-981-99-1616-0_7

more precise, rapid, and controllable power transmission with fewer losses [1]. In addition, HVDC cables are crucial to the distribution of power when it is transmitted over long distances using the offshore mode [2]. High dielectric constant polymeric materials are commonly used in the construction of HVDC cable joints and terminators, which play a crucial role in relieving the strain that has accumulated during the course of the power system's operation. In undersea power transmission systems, these cable accessories are frequently submerged in the water.

Immersion in water can cause the insulation to degrade because moisture can seep into the bulk through the polymeric chains' porous links. Even though the insulation covering the live conductor does not contribute anything to the process of transferring energy, it is just as important to the transmission and distribution of power as the live conductor itself. Polymeric materials are gaining prominence as insulating materials in long-distance electrical power transmission systems due to their technical advantages over ceramic-based insulation. Epoxy resin is an example of a thermosetting polymeric insulating material that serves to shield electrical and power electronic components from moisture and dust whilst also providing structural support for the active components [3]. For the purpose of enhancing their dielectric properties, Huang et al. have explored the use of aluminium polyethylene nanocomposites in cable joints and terminator applications [4]. The previous research shows that the insulating properties of the base polymer are maintained during the controlled addition of conductive Al nanoparticles up to a limit called the percolation threshold, all whilst providing improved dielectric, mechanical, thermal, and enhanced charge detrapping properties [5, 6].

After a certain amount of time in service, the stresses that have accumulated on the insulator's surface become too great to withstand. In addition, insulators near the coast are more likely to have conductive channels form on their surface due to the accumulation of sea salt [7]. However, when wetting agents such as humidity, fog, dew, drizzling, or misty conditions are present, the contamination layer on the surface becomes a significant and serious threat to the entire power system network. After some time, the insulating material's surface will become covered in dry bands due to the intense heating caused by the evaporation of water along the pollution-initiated conductive channels [8]. This is because the bridging eventually causes the leakage current to flow from the high voltage end to the ground electrode. The dry band's network of partial discharges can cause a total flashover, which in turn can cause the power grid to fail unexpectedly and prematurely. Thus, it is crucial to conduct laboratory research on the surface properties of the insulating material with a water droplet situated on it. Water-diffused insulating materials must also be tested in a virtual environment to determine how they will react to charging and discharging currents before being implemented in real-power networks.

As a result, the focus of this study is on the characteristics of the nanocomposite's surface after it has been subjected to water diffusion. Experiments were done to learn how water-diffused (aged) Al–epoxy nanocomposites differ from their unaged counterparts, specifically how water absorption modifies the surface potential properties of nanocomposites.

2 Materials and Method of Sample Preparation

It is clear that the overall properties of nanocomposite materials are heavily influenced by the controlled inclusion of nanofiller. That is why the percolation threshold idea is used to establish the nanofiller concentration before any composites are made. Based on the previous research, it has been determined that, whilst adding nanofiller to a material can cause the formation of conductive networks, keeping the nanofiller concentration below 14 wt% preserves the insulation properties of the base epoxy polymer [9, 10]. Thus, composite materials are prepared by adding various weight percentages (wt%) of nanofiller to the base epoxy matrix, and the insulation properties of the materials are then investigated. Nanofiller concentrations of 0, 0.5, 1, 2, 5, 10, and 14 wt% are used to prepare the specimens for this study. The nanocomposites are made with compositions like bisphenol-A epoxy resin (CY205), ethanol, tri ethylene tetra amine (TETA) hardener, and spherical aluminium nanoparticles with an average particle size of 40 nm. All of the specimens are prepared using the same protocol, which includes steps like preheating the nanoparticles in an ultrasonic bath, shear mixing, ultra-sonication, degassing, and curing. It took about 40 h of continuous inspection to make sure that each specimen was prepared under ideal conditions, and great care was taken to ensure that each composite specimen was evenly distributed. Preparation entails a number of distinct steps, each of which is described in greater depth in the aforementioned works [9, 11].

3 Experimental Setup

3.1 Water Diffusion Studies

After immersing the nanocomposite sample in water, the amount of water that diffused into the sample can be calculated by weighing the sample again. In the present study, nanocomposite specimens measuring 40×40 mm were dipped in deionized water at room temperature. At predetermined intervals, the weight of each specimen is recorded using a weighing balance accurate to within 0.01 mg. The diffusion coefficient is calculated from a graph showing the percentage increase in weight over time for a given specimen. After drying the water-diffused nanocomposites at room temperature for 24 h, they were subjected to a number of characterisation methods.

3.2 Surface Potential Measurements

In this work, the needle plane configuration setup depicted in Fig. 1 is used to investigate the charge accumulation decay behaviour of both aged and water-diffused

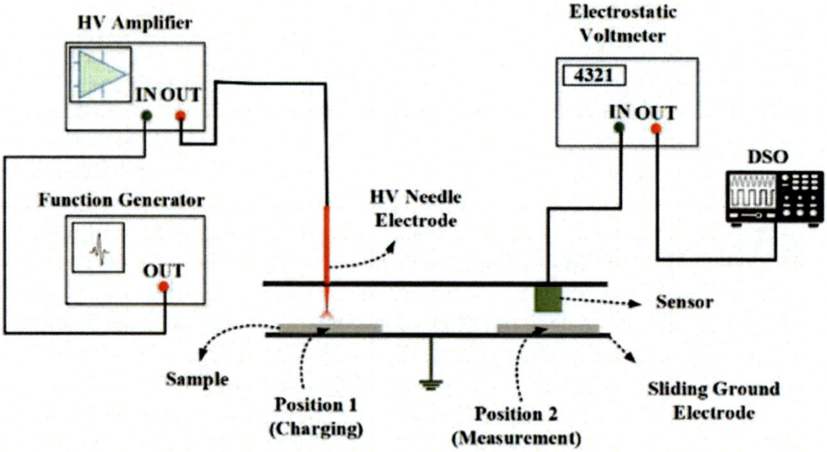

Fig. 1 Experimental setup for surface potential measurement

nanocomposites. This setup is a common non-contact and non-destructive method for measuring surface potentials. The needle electrode, whose lead is connected to the function generator and HV amplifier unit, is used to inject the charges above the test specimen (which is maintained at position 1). In this experiment, we observe the surface charge decay behaviour by applying a DC voltage of 10 kV to the needle electrode. The electrode needle's tip diameter is held constant at 0.3 mm. For three minutes, the charges will be allowed to settle over the specimen. Once the electrostatic voltmeter has been charged, the specimen is transferred (to position 2) directly under the Kelvin probe sensor. DSOs keep track of the measured potential over the specimen as a result of charge dissipation over time. The information is also used to calculate the initial potential, decay rate, mean life time, and trap depth, amongst other trap parameters.

4 Results and Discussion

4.1 Water Diffusion Studies

Figure 2 depicts the percentage weight gain of nanocomposite samples as a function of time whilst immersed in water. Regardless of the concentration of the specimen's filler, water diffusion was quick at first but slowed over time. The water uptake of each nanocomposite sample is compared using the Crank-stated diffusion coefficient, i.e.

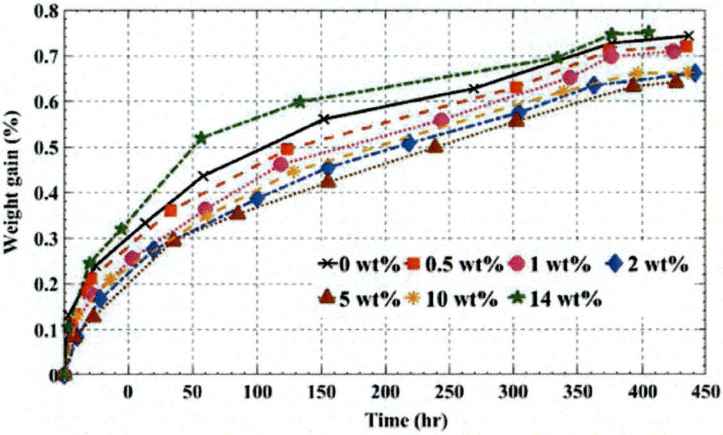

Fig. 2 Variation in weight gain percentage of water-diffused Al–epoxy nanocomposites

Diffusion coefficient,

$$D = \frac{\pi L^2}{64 t_{0.5}} \tag{1}$$

where L is the thickness of the sample and $t_{0.5}$ is the time at which the specimen's weight gain has stabilised at 50% of its steady-state value. A comparison of the diffusion coefficients of a number of nanocomposite samples is presented in Fig. 3. Water diffusion into the bulk is seen to be slightly reduced up to 5 wt% of nanofiller but is seen to increase with additions of nanofiller above this threshold. The previous research has shown that the addition of nanofiller up to 5 wt% improves interfacial adhesion between the polymer matrix and nanofillers [9]. Therefore, water diffusion is slightly reduced for nanocomposites with filler concentrations below 5 wt% [12]. This is because the available free volume is lower for these materials. In contrast, the diffusion coefficient is higher for 10 and 14 wt% specimens because of the increased available free volume caused by agglomerations for the higher filler concentration.

4.2 Surface Potential Measurements

After the charging process, the specimen is moved under a kelvin probe to record any residual potential on its surface using DSO. Since trapped carriers can dissipate without an external field, gas neutralisation is the preferred mode of detrapping the trapped charge from its localised states. Figure 4 depicts the surface potential decay characteristics of water-diffused specimens. Regardless of the polarity of the applied

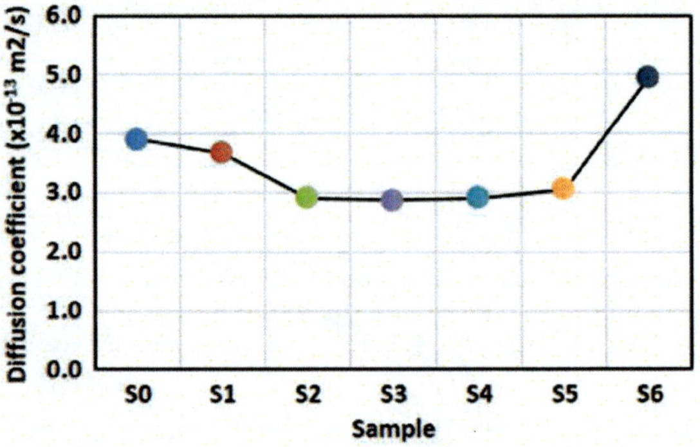

Fig. 3 Effect of water diffusion on the diffusion coefficient of Al–epoxy nanocomposites

voltage or the concentration of the filler, it is clear that the charge is decreasing at an exponential rate.

It is also possible to mathematically fit the decaying behaviour with an exponentially decaying function, i.e.

$$V(t) = V_0 * e^{-\frac{t}{\tau}} \tag{2}$$

where V_0 is the initial potential and τ is the mean life time of the trapped charge. Table 1 shows V_0 and τ for both unaged and water-aged composites for both positive and negative polarities. Diffusion of water through composites allows for the reduction of the initial potential to be observed. The injected charges may have spread through the bulk in the aged specimens, allowing them to reach the opposite ground electrode, which explains why there is less charge accumulation at the surface. When compared to the voltage profiles of unaged specimens, however, the average lifetime of trapped

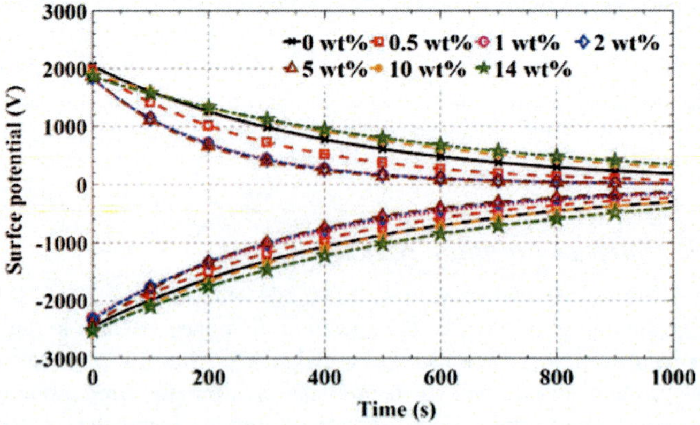

Fig. 4 Surface potential decay characteristics of Al–epoxy nanocomposites exposed to +DC and −DC voltages vary in response to water diffusion

Table 1 Measurements of the surface potential characteristics for unaged and water-diffused nanocomposites at +DC and −DC voltage profiles

Sample wt%	Unaged						Water aged					
	+DC			−DC			+DC			−DC		
	V_{in}	τ	ΔE	V_{in}	τ	ΔE	V_{in}	τ	ΔE	V_{in}	τ	ΔE
0	3860	362.71	0.87	−4634	233.81	0.86	2042	411.32	0.87	−2452	484.74	0.87
0.5	3750	265.82	0.86	−4500	226.19	0.85	1984	301.44	0.86	−2381	427.51	0.87
1	3483	186.85	0.85	−4339	218.15	0.85	1843	211.88	0.85	−2296	380.89	0.87
2	3512	163.40	0.85	−4421	177.81	0.85	1858	204.21	0.85	−2339	355.05	0.87
5	3542	131.73	0.84	−4630	127.03	0.84	1874	192.53	0.85	−2450	336.06	0.86
10	3665	468.16	0.87	−4817	250.00	0.86	1939	530.90	0.88	−2549	472.50	0.87
14	3521	521.38	0.88	−4765	294.12	0.86	1863	591.24	0.88	−2521	555.88	0.88

charges is significantly longer. The mean lifetime of aged composites is greater than that of their unaged counterparts because diffused water molecules in the aged specimens can trap injected charge carriers tightly in the absence of applied field conditions, making the detrapping of carriers difficult from their localised states. To understand how strongly the injected charges are trapped in the localised states of the nanocomposite material, a distribution analysis of the traps is performed. As the trap depth (E) varies, the density of trapped charges ($N(E)$) at various energy levels (E) mimics the decay behaviour within the bulk of the material [13, 14], i.e.

$$N(E) \propto \left| t * \frac{\mathrm{d}V}{\mathrm{d}t} \right| \tag{3}$$

and trap depth,

$$\Delta E = E_c - E_d = kT * \ln(vt) \tag{4}$$

In this equation, v is the attempt frequency, on the order of 10^{12} s^{-1}, and E_c is the lowest energy level of the transport state (separating the empty and filled states). Figure 5 depicts the resulting variation in trap characteristics of Al–epoxy nanocomposites upon water diffusion. Trap depth, defined as the energy required for a trapped carrier to escape its localised state, is found to be greater in water-diffused nanocomposites than in their unaged counterparts. Water ingress causes charge carriers to become trapped, which requires more energy than in unaged composites. Table 1 displays the trap depth values of both unaged and water-aged nanocomposites for a variety of voltage profiles. There was a direct correlation between mean life time and trap depth, with a higher mean life time resulting in a deeper trap, and vice versa.

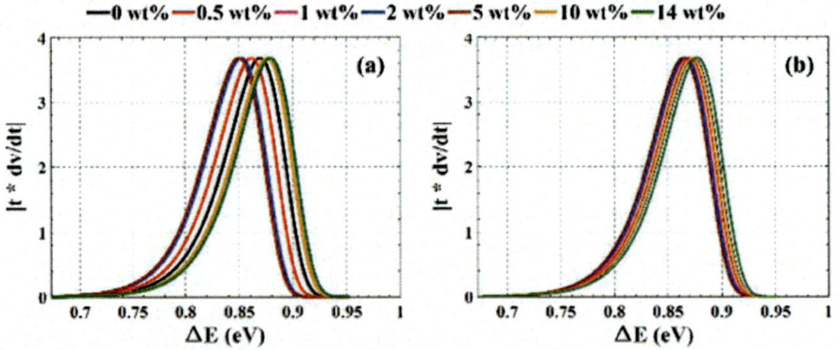

Fig. 5 Trap distribution characteristics of water-diffused Al–epoxy nanocomposites

5 Conclusion

In this study, the surface charge characteristics of unaged and water-diffused Al–epoxy nanocomposite samples were analysed for its HVDC insulation application. The study found that by adding nanoparticles to an epoxy polymer matrix in a controlled manner (up to 5 wt%), the available free volume is reduced, thereby decreasing the amount of water ingress into the composite material as a whole. The surface potential has decayed exponentially over time, regardless of filler concentration, voltage polarity, or ageing technique. Epoxy nanocomposites that were subjected to water diffusion saw a decrease in initial potential and an increase in mean life time compared to their unaged counterparts. For any given sample, there is a linear correlation between mean lifetime and trap depth was noticed.

References

1. Chen G, Hao M, Xu Z, Vaughan A, Cao J, Wang H (2015) Review of high voltage direct current cables. CSEE J Power Energy Syst 1(2):9–21
2. Du BX, Ran ZY, Li J, Liang HC, Yao H (2020) Fluorinated epoxy insulator with interfacial conductivity graded material for HVDC gaseous insulated pipeline. IEEE Trans Dielectr Electr Insul 27(4):1305–1312
3. Du BX, Wang MY, Li J, Xing YQ (2018) Temperature dependent surface charge and discharge behavior of epoxy/AlN nanocomposites. IEEE Trans Dielectr Electr Insul 25(4):1300–1307
4. Huang XY, Jiang PK, Kim CU (2007) Electrical properties of polyethylene/aluminum nanocomposites. J Appl Phys 102(12):124103
5. Naresh C, Jayaganthan R, Danikas MG, Tanaka T, Sarathi R (2019) Understanding the dielectric and mechanical properties of self-passivated Al–epoxy nanocomposites. IET Sci Meas Technol 13(9):1336–1344
6. Chillu N, Jeshurun A, Jayaganthan R, Velmurugan R, Sarathi R (2019) Investigation on dielectric and mechanical properties of epoxy reinforced with glass fiber and nano-silica composites. Mater Res Express 6(11):115082

7. Vinod P, Desai BMA, Sarathi R, Kornhuber S (2019) Investigation on the electrical, thermal and mechanical properties of silicone rubber nanocomposites. IEEE Trans Dielectr Electr Insul 26(6):1876–1884
8. Zhu Y (2019) Influence of corona discharge on hydrophobicity of silicone rubber used for outdoor insulation. Polym Test 74:14–20
9. Chillu N, Jayaganthan R, Nageshwar Rao B, Danikas M, Tanaka T, Sarathi R (2020) Investigation on space charge and charge trap characteristics of Al–epoxy nanocomposites. IET Sci Meas Technol 14(2):146–156
10. Qiang D, Wang Y, Wang X, Chen G, Andritsch T (2019) The effect of filler loading ratios and moisture on DC conductivity and space charge behaviour of SiO_2 and hBN filled epoxy nanocomposites. J Phys D Appl Phys 52(39):395502
11. Chillu N, Sarathi R, Jayaganthan R (2021) Impact of gamma-irradiation on space charge and charge trap characteristics of Al/epoxy nanocomposites. SPE Polym 2(1):38–49
12. Babu MS, Sarathi R, Vasa NJ, Imai T (2020) Investigation on space charge and charge trap characteristics of gamma-irradiated epoxy micro–nano composites. High Volt 5(2):191–201
13. Gao Y, Li N, Li J, Du B, Liu Z (2020) Charge transport behavior in gamma-ray irradiated poly(ethylene terephthalate) estimated by surface potential decay. High Volt 1–13
14. Li J, Liang HC, Du BX, Wang ZH (2019) Surface functional graded spacer for compact HVDC gaseous insulated system. IEEE Trans Dielectr Electr Insul

A Study on Impact of Hydrophobic Effect on Al_2O_3 Coated Glass by Sol–Gel Dip Coating Method for Automobile Windshield Application

K. Chandru and R. Elansezhian

Abstract This work aimed to achieve hydrophobic effect on aluminum based on nano/micro hierarchical surface structure through sol–gel coating technique. A hydrophobic coating for glass substrates was developed using Tetraethyl orthosilicate $Si(OC_2H_5)_4$ synthesized with Al_2O_3 nanoparticles. The sol was prepared by adding the Al_2O_3 nanoparticles to sol just before the coating process and dispersing. The coating process was conducted using a sol–gel dip coating deposition method. The effect of functionalization of Al_2O_3 nanoparticles to prepare the highly dispersed nanoparticles has been investigated in detail. The surface morphology was characterized by field emission scanning electron microscopy (FESEM), and surface roughness were measured. The effect of Al_2O_3 on the hydrophobicity of the coating was evaluated using contact angle measurements. The results reveal that water contact angles improved with the addition of Al_2O_3 nanoparticle wt% 0.5, 1, and 1.5 g. The glass coated with higher wt% of Al_2O_3 nanoparticle exhibits a water contact angle of 113° with an effect to lower surface roughness value of 0.011 μm. A water repelling coating on the glass is developed with the incorporation of Al_2O_3 nanofiller via sol–gel synthesis method.

Keywords Hydrophobic · Tetraethyl orthosilicate · Water contact angle · Al_2O_3 · Sol–gel

1 Introduction

Functional qualities like water repellency, non-wettability, and self-cleaning offer value to industrial products including clothing, metal sheets, and glass panels [1]. Numerous researchers have produced hydrophobic coatings to provide useful qualities to glasses as well as other materials. Reduced surface energy is known to result in hydrophobic surfaces. Greater than ninety-degree static contact with the angles of a liquid droplet sitting on a glass surface are characteristic of hydrophobic nature.

K. Chandru · R. Elansezhian (✉)
Department of Mechanical Engineering, Puducherry Technological University, East Coast Road, Pillaichavadi, Puducherry 605014, India
e-mail: elansezhianr@ptuniv.edu.in

N. M. Rao et al. (eds.), *Advanced Nanomaterials and Their Applications*,
Springer Proceedings in Materials 22, https://doi.org/10.1007/978-981-99-1616-0_8

Water-repellent coatings that are capable of resisting water have garnered a great deal of interest due to their broad range of technical uses, including vehicle glass, construction materials, bathroom glass, and greenhouse glass. Automobile glass has several essential uses for water-repellent coatings. They increase a driver's ability to see properly and drive more securely by reducing glare inside the rain at nighttime, since a very hydrophobic surface causes droplets to bead and roll off [4]. This work explored the surface functionalization of glass for the formation of water-repellent glass and assessed the physical as well as surface features of water-repellent coatings. The sol–gel method was used to produce water-repellent coatings on the surfaces of glass. In a similar manner, Al_2O_3 nanoparticles were added to the nanocomposite layer produced by the reactive chemical etching approach using tetraethyl orthosilicate and Glycidyloxypropyltrimethoxysilane. The impact of functionalization of highly scattered Al_2O_3 nanoparticles inside a silane layer was examined 168° contact angle attained [2].

2 Experimental Section

2.1 *Materials and Methods*

The chemicals TEOS tetraethyl orthosilicate, Al_2O_3 nanopowders, ethanol, and acetone are procured from Sai Scientific Pvt. Ltd, India. A magnetic stirrer has been used to mix the chemical at a specified rate of speed. Initially, 30 × 30 cm glass was procured from the TPRS Company, Puducherry, India. The company was well known for windshield glass manufacturing. This type of glasses are commonly used around the automobile industries to structure the windshields. Using diamond cutter, required glass samples were extracted from that glass. Totally 20 samples are taking out from it. A first set of 10 samples are in size of 40 × 20 mm.

The samples were chemically treated with three solutions such as distilled water, ethanol, and acetone to avoid the deposit of unwanted foreign nanoparticles and dried at ambient temperature to make a surface ready for coating (Fig. 1).

A 30 ml of tetraethyl orthosilicate is mixed with 31 ml ethanol and has stirred for about 5 min and a solution of 38 ml of distilled water plus 3–4 drops of HCL +60 °C then Al_2O_3 was added with different wt% 0.5, 1, 1.5 was treated up to 1.5 h with steady stirring and as a final stage the formation of clear white precipitate (TEOS sol–gel) is formed from soluble state to gel state (Fig. 2).

2.2 *Dip Coating Process*

Dipping of samples in solvent at the withdrawal rate of 15 mm/s min. The samples underwent calcination process for 90 min at 400 °C in muffle furnace to remove the

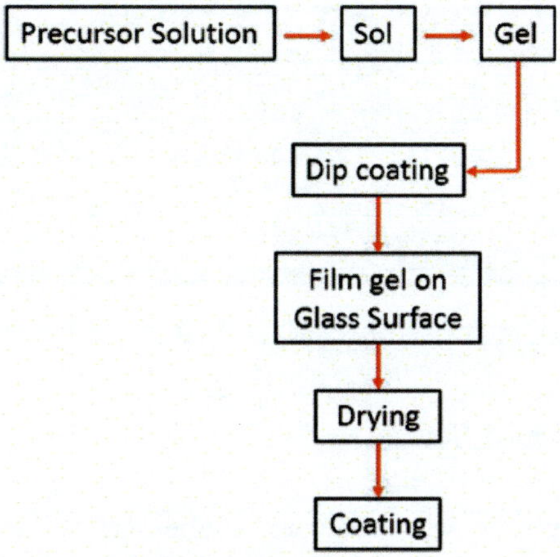

Fig. 1 Preparation of coatings by sol–gel method

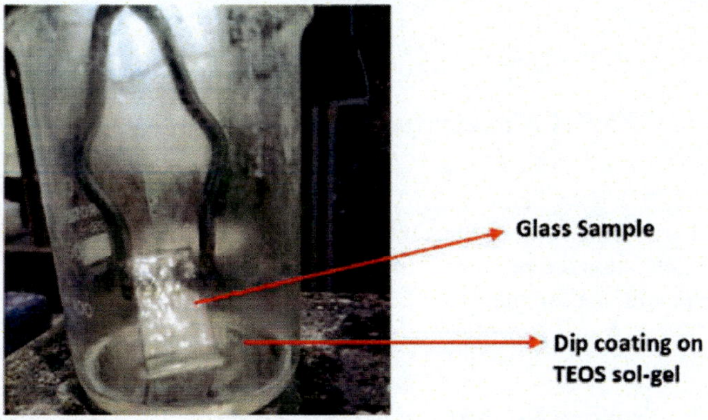

Fig. 2 Coating of samples

xerogel from the sol–gel coating which allows the coating to stagnate on the sample surface (Fig. 3).

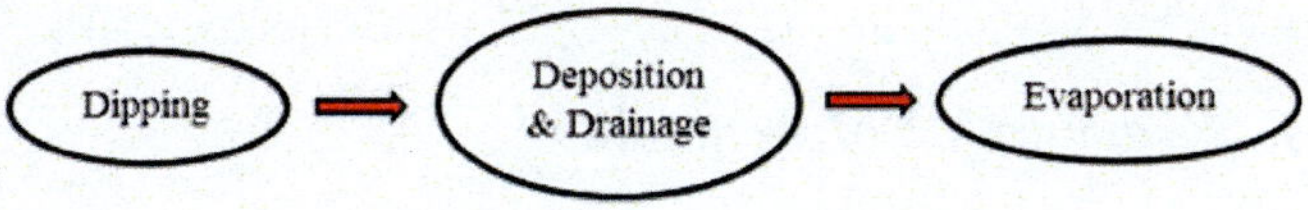

Fig. 3 Stages of dip coating method

Fig. 4 **a–c** Al_2O_3 FESEM micrographs [1, 5 μm]

3 Results and Section

A liquid droplet is deposited onto a solid surface, it will form a contact angle depending on the surface tension of the solid and the interfacial tension between solid and liquid. The angle created by the solid surface and the tangent of the droplet is known as the "contact angle," and this equation is also known as the Young equation [24].

3.1 FESEM Micrographs

Al_2O_3 1.5 g wt below FESEM micrographs Fig. 4a–c at the magnification with 1, 5 μm which shows the nanoparticles of Al_2O_3, and the samples with different magnification are presented. In 1 μm, deposition of coating confirms the pattern of coating on the sample.

3.2 Surface Roughness Measurement

The surface roughness of the coated samples were analyzed using Mitutoyo surface roughness tester. The sample with 1.5 g Al_2O_3 demonstrates the maximum water contact angle of 113°. Coated sample tends to develop a hydrophobic effect with lower surface roughness value of 0.011 μm (Fig. 5).

3.3 Examination of Wettability

Wetting fluid parameter are as mentioned below

Wetting fluid used on the surfaces: Water

Height of the surface from droplet source: 5 cm

Wetting time of the surfaces: 60 s

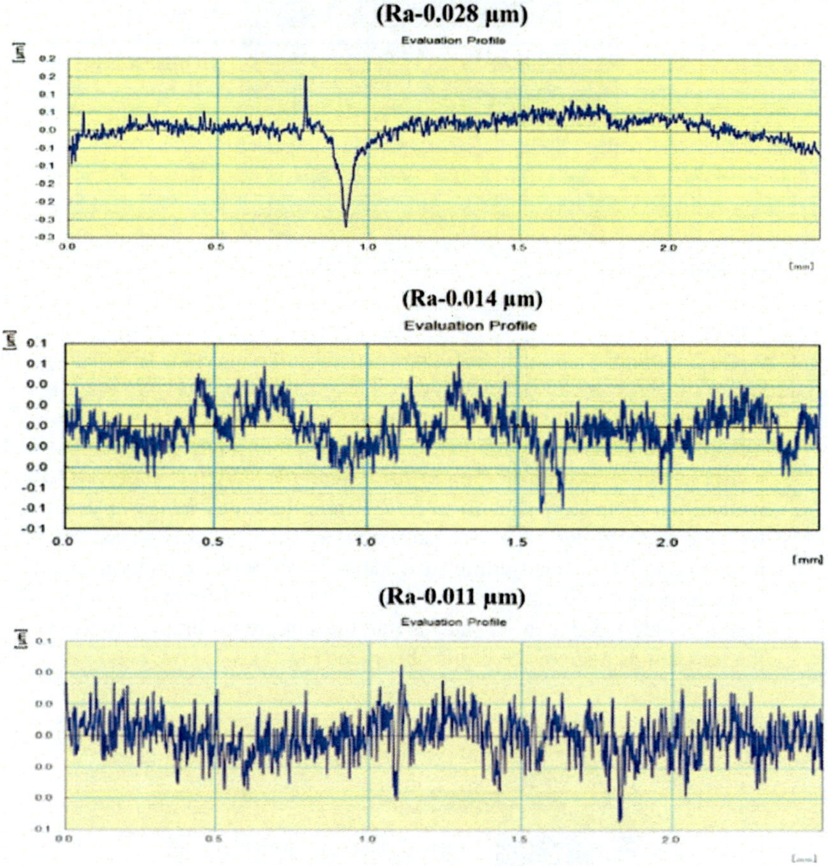

Fig. 5 **a–c** surface roughness measurement of Al_2O_3 0.5, 1, 1.5 g wt%

Inclination of surface with the horizontal: 0°.

In wettability testing, an uncoated sample sessile water droplet attaches to its surface with increased surface tension and draws water, whereas in case of coated sample causes the water droplet to form into a mounted form because coating reduces adhesion and surface energy when compared to an uncoated sample (Fig. 6).

3.4 *Water Contact Angle Test*

The WCA water contact angle test has been performed with Optical Tensiometer (Theta Lite model) Program—One Attention version 2.4 Standard—ASTM D7334-08. Optical tensiometer is used to calculate the angle between two bodies. In surface

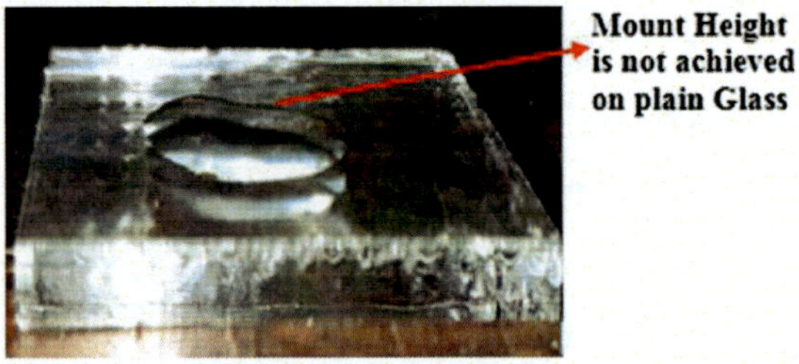

Fig. 6 Wetting of plain glass

science, the contact angle of a droplet deposited on a surface can be measured using an optical tensiometer and a microscope. The contact angle is calculated using modern technology that includes a light source, camera, and analytical software to figure out a droplet's contact angle in order to assess a sample's wettability. The wettability of a sample measures the absorbability of the sample [25]. It is important to understand wettability of surfaces such as polyethylene, glass, or fabric in order to know how well the material will repel or absorb a liquid such as water. When contact angle is measured to be 0°, this signifies that the wettability of the material is absolute. The material completely absorbed the drop of liquid. A droplet on a sample that has a contact angle more than 90° means that the sample is completely hydrophobic and the meaning is none of the liquid absorbed (Fig. 7).

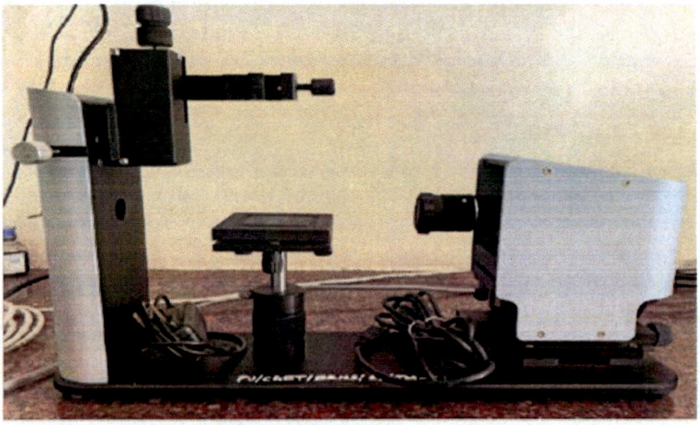

Fig. 7 Optical tensiometer (Theta Lite model)

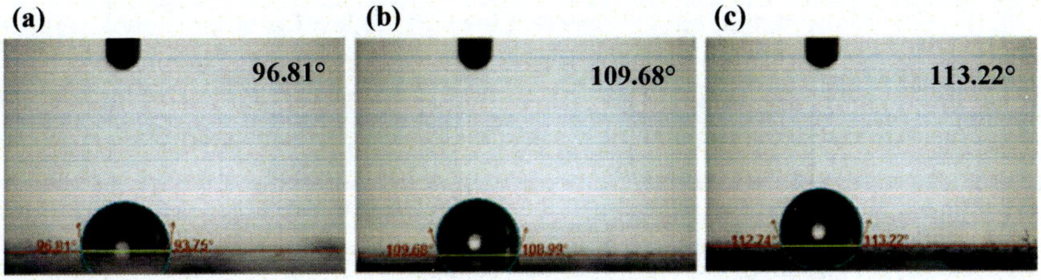

Fig. 8 **a–c** Al_2O_3 (0.5 g), Al_2O_3 (1 g), Al_2O_3 (1.5 g) coated sample

3.5 Hydrophobic Test Results

The test results presented below are the water contact angle images observed from the optical tensiometer instrument and the different images contains 0.5, 1, and 1.5 g weightage of the Al_2O_3 nanoparticle coated with sol–gel TEOS synthesis. Each observed and showed different contact angle, the 0.5 g produces a contact angle of 93.75–96.81°, and 1 g coated presents an angle of 108.99–109.68° as a sample contains maximum ratio of Al_2O_3 coated 1.5 g shows a contact angle of 112.74–113.22° (Figs. 8 and 9).

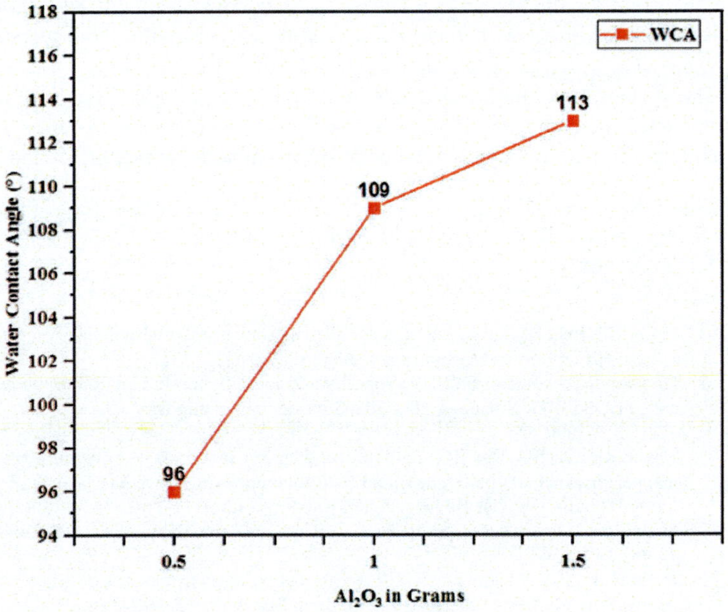

Fig. 9 Water contact angle measurement graph

The hydrophobic effect created at the surface level is proportional to surface free energy, depending on the wt% of nanoparticle of Al_2O_3 0.5, 1, and 1.5 g. The graphs show that as the weight ratio of Al_2O_3 rises, the surface energy and adhesion of the material decrease, and the water contact angle rises proportionately, whereas for 0.5 g a lower contact angle of 93° and 1.5 g for a higher water contact angle of 113° has been recorded.

The surface modified after coating will be smoother than the unmodified surface before coating. More hydrophobicity was attained with low surface roughness when compared to two other cases, the least surface roughness value for the 1.5 g wt% Al_2O_3 is 0.011 μm. Therefore, as surface roughness decreases, surface energy and adhesion decrease which makes the water droplet exhibits higher water contact angle.

4 Conclusion

Thus, the study on hydrophobic effect for the application of automobile windshield glass using the sol–gel technique was successfully coated on the glass using the dip coating method. The physical characteristics and the change in the surface of the coated area have been characterized using FESEM. With an increase in the wt% of Al_2O_3, the surface energy drops, improving the water contact angle while concurrently reducing surface adhesion. The Al_2O_3 1.5 g coated sample exhibits a hydrophobic effect with water contact angle of 113° with a low surface roughness value of 0.011 μm. The water contact angle improved with 30.19% to 35.53%. The surface roughness reduced with 52.83% to 20.75%. Thus improved surface roughness percentage is 32.08%. This will increase glass transparency while raining, and additionally it may effectively produces an anti-fogging effect and avoids mist.

References

1. Latthe SS, Imai H, Ganesan V, Rao AV (2010) Porous superhydrophobic silica films by sol–gel process. Microporous Mesoporous Mater 130(1–3):115–121
2. Khodaei M, Shadmani S (2019) Superhydrophobicity on aluminum through reactive-etching and TEOS/GPTMS/nano-Al_2O_3 silane-based nanocomposite coating. Surf Coat Technol 374:1078–1090
3. Nayak RK, Dash A, Ray BC (2014) Effect of epoxy modifiers (Al_2O_3/SiO_2/TiO_2) on mechanical performance of epoxy/glass fiber hybrid composites. Procedia Mater Sci 6:1359–1364
4. Jeong HJ, Kim DK, Lee SB, Kwon SH, Kadono K (2001) Preparation of water-repellent glass by sol–gel process using perfluoroalkylsilane and tetraethoxysilane. J Colloid Interface Sci 235(1):130–134
5. Ikeda H, Fujino S, Kajiwara T (2011) Fabrication of micropatterns on silica glass by a room-temperature imprinting method. J Am Ceram Soc 94(8):2319–2322
6. Zhang Y, Wu Y, Chen M, Wu L (2010) Fabrication method of TiO_2–SiO_2 hybrid capsules and their UV-protective property. Colloids Surf, A 353(2–3):216–225
7. Yuan F, Peng H, Yin Y, Chunlei Y, Ryu H (2005) Preparation of zinc oxide nanoparticles coated with homogeneous Al_2O_3 layer. Mater Sci Eng, B 122(1):55–60

8. Dai J, Gao W, Liu B, Cao X, Tao T, Xie Z, Zhao H, Chen D, Ping H, Zhang R (2016) Design and fabrication of UV band-pass filters based on SiO_2/Si_3N_4 dielectric distributed Bragg reflectors. Appl Surf Sci 364:886–891
9. Pana I, Vitelaru C, Kiss A, Zoita NC, Dinu M, Braic M (2017) Design, fabrication and characterization of $TiO_2–SiO_2$ multilayer with tailored color glazing for thermal solar collectors. Mater Des 130:275–284
10. Jiang Z, Fang S, Wang C, Wang H, Ji C (2016) Durable polyorganosiloxane superhydrophobic films with a hierarchical structure by sol–gel and heat treatment method. Appl Surf Sci 390:993–1001
11. Parale VG, Mahadik DB, Kavale MS, Mahadik SA, Rao AV, Mullens S (2013) Sol–gel preparation of PTMS modified hydrophobic and transparent silica coatings. J Porous Mater 20(4):733–739
12. Dhere SL, Latthe SS, Kappenstein C, Pajonk GM, Ganesan V, Rao AV, Wagh PB, Gupta SC (2016) Transparent water repellent silica films by sol–gel process. Appl Surf Sci 256(11):3624–3629
13. Ganbavle VV, Bangi UK, Latthe SS, Mahadik SA, Rao AV (2011) Self-cleaning silica coatings on glass by single step sol–gel route. Surf Coat Technol 205(23–24):5338–5344
14. Zuo Z, Gao J, Liao R, Zhao X, Yuan Y (2019) A novel and facile way to fabricate transparent superhydrophobic film on glass with self-cleaning and stability. Mater Lett 239:48–51
15. Huang X, Tepylo N, Pommier-Budinger V, Budinger M, Bonaccurso E, Villedieu P, Bennani L (2019) A survey of icephobic coatings and their potential use in a hybrid coating/active ice protection system for aerospace applications. Prog Aerosp Sci 105:74–97
16. Zhu T, Cheng Y, Huang J, Xiong J, Ge M, Mao J, Liu Z, Dong X, Chen Z, Lai Y (2020) A transparent superhydrophobic coating with mechanochemical robustness for anti-icing, photocatalysis and self-cleaning. Chem Eng J 399:125746
17. Dhere SL, Bangi UK, Latthe SS, Rao AV (2011) Enhancement in hydrophobicity of silica films using metal acetylacetonate and heat treatment. J Phys Chem Solids 72(1):45–49
18. Satapathy M, Varshney P, Nanda D, Mohapatra SS, Behera A, Kumar A (2018) Fabrication of durable porous and non-porous superhydrophobic LLDPE/SiO_2 nanoparticles coatings with excellent self-cleaning property. Surf Coat Technol 341:31–39
19. Latthe SS, Imai H, Ganesan V, Rao AV (2009) Superhydrophobic silica films by sol–gel coprecursor method. Appl Surf Sci 256(1):217–222
20. Skroznikova V, Popovich N, Dimitrov T (2013) Transparent hydrophobic sol–gel silica coatings on glass. Sci Works Russ Univ 52:61–64
21. Fujiwara M, Imura T (2015) Transparent silica thin films prepared from sodium silicate and bovine serum albumin with petal effect. Ceram Int 41(6):7565–7572
22. Widati AA, Nuryono N, Kartini I (2019) Water-repellent glass coated with $SiO_2–TiO_2$ methyltrimethoxysilane through sol–gel coating. AIMS Mater Sci 6(1):10–24
23. Sanchez C, Belleville P, Popall M, Nicole L (2011) Hybrid materials themed issue. Chem Soc Rev 40:453–1152
24. Li D, Neumann AW (1992) Contact angles on hydrophobic solid surfaces and their interpretation. J Colloid Interface Sci 148(1):190–200
25. Khorsand S, Raeissi K, Ashrafizadeh F, Arenas MA (2015) Super-hydrophobic nickel–cobalt alloy coating with micro-nano flower-like structure. Chem Eng J 273:638–646

Design and Analysis of Chalcogenide GeAsSe Waveguide for Dispersion Properties

V. Hitaishi, K. Jayakrishnan, and Nandam Ashok

Abstract This paper reports the design and analysis of a $Ge_{11.5}As_{24}Se_{64.5}$ chalcogenide optical waveguide. The structure consists of $Ge_{11.5}As_{24}Se_{64.5}$ as a core material and $Ge_{11.5}As_{24}S_{64.5}$ is considered as cladding material. Dispersion, mode, profile, and propagation loss analysis of the waveguide are considered in the near and mid-infrared spectral regions. The designed structure reports a −37.96 ps/nm km at 3 mm wavelength and 0.716 ps/nm km dispersion at 4 mm wavelength. The propagation loss of fundamental mode is 1.84, 2.20, and 1.97 dB/cm at 3, 4, and 6 mm wavelengths, respectively. These results show that the proposed optical waveguide design should find applications in supercontinuum generation.

Keywords Dispersion · Mode fields · Optical waveguides · Supercontinuum generation

1 Introduction

In general, any structure that helps the flow of an electromagnetic wave through it by confining it within itself is called an optical waveguide. An optical waveguide can be considered as a building block of any optical circuit [1, 2]. A linearly elongated high-index medium called core is surrounded by a low-index medium called cladding, constitutes an optical waveguide. It helps us interconnect the optical components and enables the propagation of light at an extremely lower loss. These optical waveguides have been widely classified based on their geometry, refractive index, propagation of mode, and material [3–5]. These waveguides result in different responses for linear and nonlinear phenomena [6]. Over the years, waveguides have been extensively used in various applications such as sensors [7, 8], resonators [9, 10], filters [11], and certain nonlinear applications such as supercontinuum generation (SCG) [12], wavelength conversion [13], soliton formation [14], four-wave mixing [15], and frequency comb generations [16]. Each of these applications is

V. Hitaishi · K. Jayakrishnan · N. Ashok (✉)
Department of Physics, School of Advanced Sciences, VIT-AP University, Amaravati, Andhra Pradesh 522237, India
e-mail: nandam.ashok@gmail.com

N. M. Rao et al. (eds.), *Advanced Nanomaterials and Their Applications*,
Springer Proceedings in Materials 22, https://doi.org/10.1007/978-981-99-1616-0_9

made possible by relying on certain properties of waveguides. Among those dispersion has a particular role in nonlinear applications. In most cases of nonlinear applications like supercontinuum generation, a large bandwidth with the ability to pump pulses in different wavelengths near zero-dispersion is recommended. Achieving a near zero flat dispersion is hard because of strong field confinement and waveguide material dispersion. Though waveguides have been classified on a different basis, the material on which the waveguide is made has a significant interest here, because it plays an imminent role in getting a low dispersion with multiple zero-dispersion wavelength (ZDW). Despite having a large pool of materials with good nonlinear properties, chalcogenide glasses paved their way in nonlinear optics due to their high non-linearity, transmission with low loss, and a broad transparency window that falls over visible and infrared wavelength ranges [17]. These chalcogenide glasses have been formed by combining chalcogenide elements such as Ge, As, or Sb. Because most chalcogenide glasses used in nonlinear optics have a large negative material dispersion in the near-infrared band, the waveguide structure must be carefully designed to achieve a proper waveguide dispersion for compensating the material dispersion [18]. Also, while designing the chalcogenide glass waveguides, both the requirement for dispersion and the feasibility of fabrication should be considered.

Ashok et al. proposed a horizontal slot-strip-slot chalcogenide waveguide structure and achieved a 0–350 ps/nm km dispersion over a bandwidth from 2512 to 3887 nm [19]. Karim et al. reported supercontinuum generation broadening over a wavelength range from 3.5 to 15 μm using an As_2Se_3 channel waveguide with 170 fs pulse pumped at 6 μm with an input peak power of 10 kW [20]. Xia et al. reported broadband supercontinuum generation covering 2–15 μm using a Ge–As–Te–Se/Ge–Se chalcogenide waveguide with 120 fs pulses pumped at 5.8 μm wavelength with an input peak power of 800 W [21]. Zhang et al. reported a slot waveguide design and studied the resonance phenomenon between the slot mode and slab waveguide mode [22]. It shows a dispersion value of −181,520 ps/nm km due to the interaction between the strip and slot structure modes. Zhu et al. have designed and analyzed the dual-slot silicon structure to study the dispersion properties, and the design shows a very low dispersion and reports the zero-dispersion wavelength at four different wavelengths [23]. Zhai et al. have investigated the ridge-slot waveguide structure by considering the chalcogenide materials, the design results in a flat dispersion with 3 zero-dispersion points, and the dispersion consists between the −26 and + 27 ps/nm km [18].

Various applications can arise with the knowledge of the dispersion properties of the waveguide, and Vidhi Mann et al. designed a coupled strip-slot waveguide for dispersion compensation. It can be used in linear pulse compressor and integrated optic dispersion compensator [24]. Jiayao Huang et al. designed a tapered silicon waveguide to demonstrate simultaneous pulse combination and self-similar pulse compression [25]. Mulong Liu et al. proposed a dispersion-flattened technology using nanophotonic waveguides to produce multiple octaves of supercontinuum generation [26].

In the present article, the designed waveguide structure consists of three slab waveguides placed one on another considered as a core region made with $Ge_{11.5}As_{24}Se_{64.5}$ chalcogenide material. The cladding region of waveguide is made with $Ge_{11.5}As_{24}S_{64.5}$ material. Calculation of the effective indices and mode fields of the waveguide was done using the COMSOL 6.0 software and the dispersion of the waveguide was calculated by using the MATLAB R2021b software. The proposed research should find applications in supercontinuum generation.

2 Proposed Waveguide Structure

The vertical cross-section of the proposed chalcogenide waveguide is shown in Fig. 1. The waveguide's core is consisting of three rectangular slabs with $Ge_{11.5}As_{24}Se_{64.5}$ material. The cladding is made of $Ge_{11.5}As_{24}S_{64.5}$ material. The structure is designed in a way that 3 slabs are stacked upon one another. The parameters of the waveguide are optimized for low loss and good dispersion over a broad wavelength. Optimized waveguide parameters are, top slab width (a) = 4 μm, middle slab width (b) = 8 μm, waveguides width and the bottom slab width (c) = 15 μm, top slab height (d) = 2.2 μm, middle slab height (e) = 2 μm, bottom slab height (f) = 1.5 μm, waveguide height (g) = 9 μm, and length from top of waveguide to bottom slab (h) = 5.55 μm.

The wavelength-dependent refractive index of the core and cladding materials is calculated by considering the Sellmeier equation of the core and cladding materials [27].

$$n_{Ge_{11.5}As_{24}Se_{64.5}}(\lambda) = \sqrt{1 + \frac{5.78525\lambda^2}{\lambda^2 - 0.28795^2} + \frac{0.39705\lambda^2}{\lambda^2 - 30.39338^2}} \tag{1}$$

$$n_{Ge_{11.5}As_{24}S_{64.5}}(\lambda) = \sqrt{1 + \frac{4.18011\lambda^2}{\lambda^2 - 0.31679^2} + \frac{0.35895\lambda^2}{\lambda^2 - 22.77018^2}} \tag{2}$$

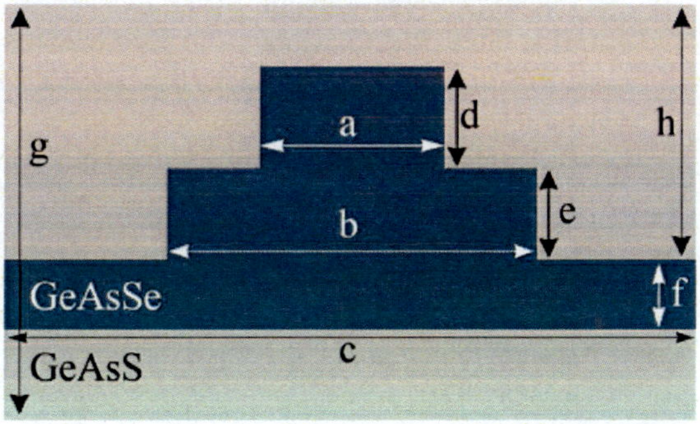

Fig. 1 Schematic of the proposed waveguide structure parameter

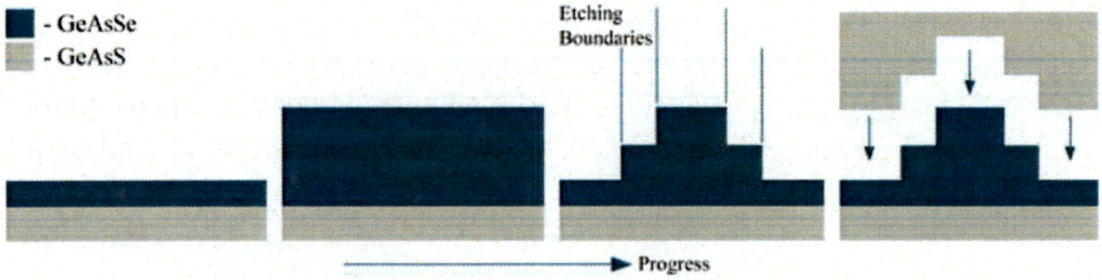

Fig. 2 Fabrication procedure

where λ is wavelength in micrometers. The wavelength-dependent refractive index helps in calculating the effective index (n_{eff}), effective area (A_{eff}), propagation loss, and mode profiles at the required wavelength.

The lower cladding material and the first bottom layer of the core can be fabricated through film deposition. Later, placing a core material followed with Inductively Coupled Plasma-Reactive Ion Etching (ICP-RIE) process to make the shape of the core. To fill the rest of the cladding material, film deposition can be done again [28]. Soft surfaces of waveguides can be obtainable using photoresist coating over the materials followed by the deep-UV lithography from ICP etching or from wet etching process during fabrication [29] (Fig. 2).

3 Numerical Results and Analysis

The present paper numerically analyzed for the following parameters, a = 4 μm, b = 8 μm, c = 15 μm, d = 2.2 μm, e = 2 μm, f = 1.5 μm, g = 9 μm, and h = 5.55 μm. The waveguide was designed by utilizing the above waveguide parameters. The effective index of the fundamental TE mode is numerically calculated for different wavelengths. Figure 3 shows the effective index of the mode as a variable of wavelength. The plotted figure shows that, as wavelength increases, the effective index of the mode decreases. Further investigation shows that the mode fields of the waveguide for various wavelengths and the results are presented in Fig. 4.

Numerical investigation of the dispersion is done by using the effective index of the mode. Figure 5 shows the dispersion curve as a variable of wavelength. The dispersion curve results in a low dispersion at 3 mm and zero-dispersion value around 4 mm wavelength. The structure shows a value of −37.96 ps/nm km for 3 mm wavelength and −0.716 ps/nm km for 4 mm wavelength. Along with this, calculation of propagation loss of the mode for different values of wavelength is considered. The loss curve of the mode is shown in Fig. 6, observing the curve that waveguide has a low loss at wavelength region below 3 μm. Then it slightly increasing around 4 μm, again the losses are lesser trailing with a large increase in loss after 7 μm wavelength range. The proposed structure shows a low loss of 1.84 dB/cm at 3 μm wavelength and 1.97 dB/cm at 6 μm wavelength. We can optimize the waveguide for

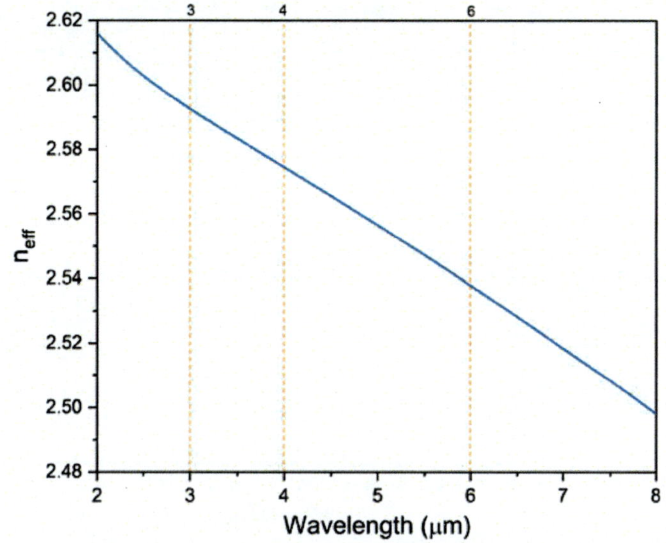

Fig. 3 Variation of fundamental mode effective index with wavelength

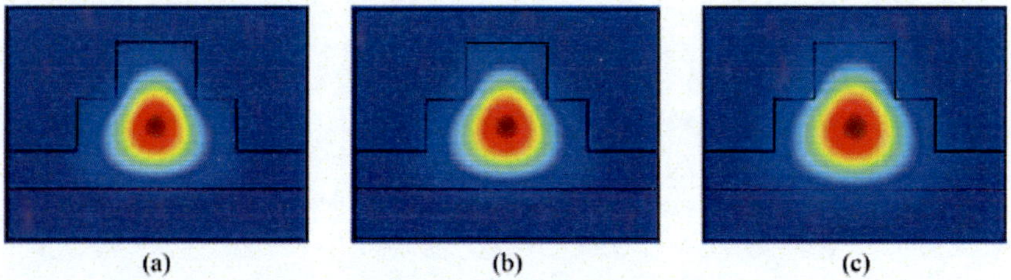

Fig. 4 Fundamental mode profiles for **a** 3 μm **b** 4 μm and **c** 6 μm wavelength

zero-dispersion around 4 and 6 μm wavelength. As a result, the proposed structure can be utilized for supercontinuum generation applications.

In the current section, we have investigated the tolerance of waveguide parameters for dispersion and loss. The presented waveguide is optimized structurally with the consideration of transmission loss and dispersion curve.

In Fig. 7a, we have plotted the variation of the dispersion curve with wavelength for different values of middle slab height 'e', and other design parameters such as a, b, c, d, f, g, and h have been kept constant. From Fig. 7a, we can observe a clean flat dispersion curve for e = 1.2 μm which spans from 5 to 6.5 μm wavelength. While increasing the value of e, the dispersion curve is moving away from the zero-dispersion line into the anomalous dispersion regime. For e = 1.2 μm, the structure shows a dispersion of −35.5 ps/nm km at 3 μm wavelength and a dispersion of 0.6 ps/nm km at 4 μm wavelength. For e = 0.4 μm, we can see that the curve is

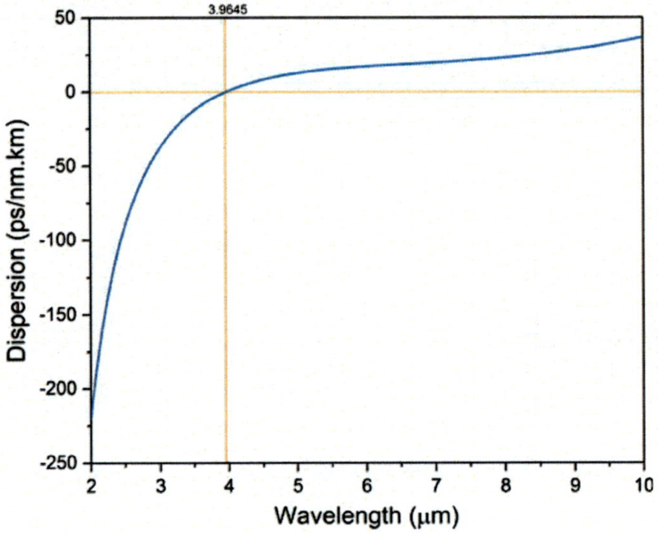

Fig. 5 Dispersion curve as a function of wavelength

Fig. 6 Loss of the mode for different values of wavelength

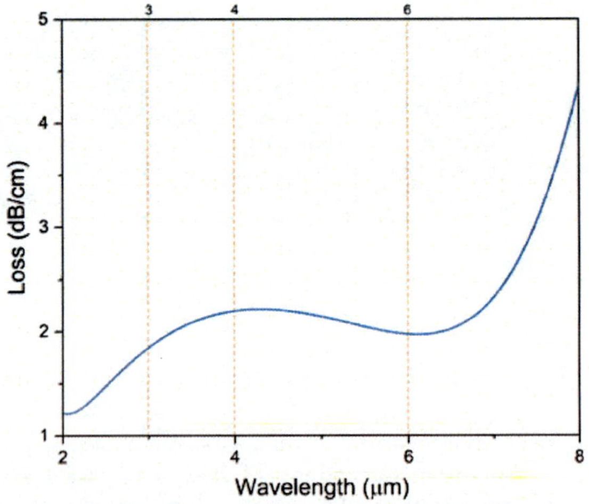

passing the zero-dispersion line at three different points. They are at 4.037, 6.136, and 8.165 μm. In addition, the author also evaluated the variation of dispersion with wavelength for the waveguide parameters 'a' and 'd'.

The dispersion curve for various values of 'a' is calculated numerically and plotted in Fig. 7b. From the calculated results, we observed that for a = 2 mm the design achieves a dispersion of −37.3 and −1.4 ps/nm km at 3 and 4 mm wavelength, respectively. For a = 3 mm, it results in a dispersion of −37.7 and −0.4 ps/nm km

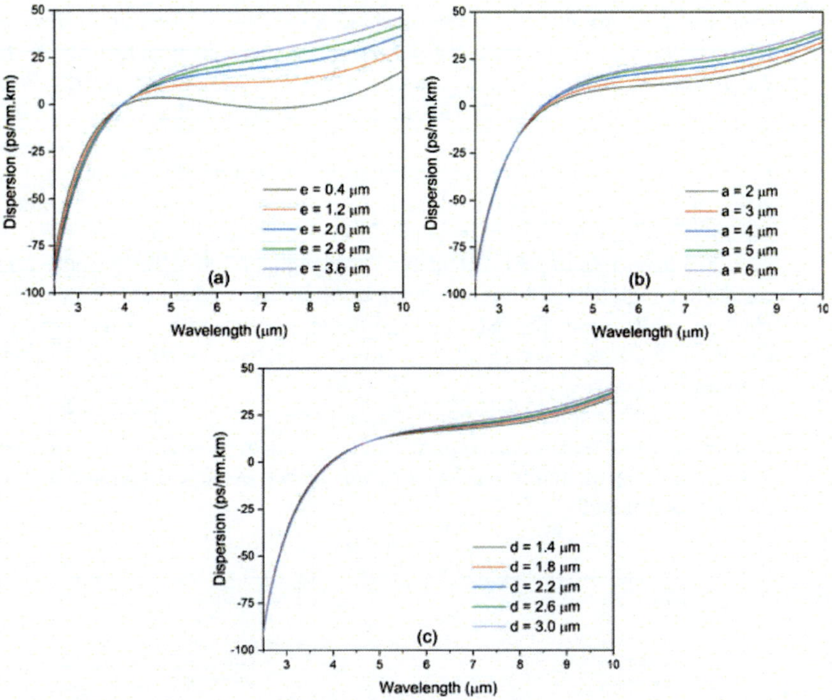

Fig. 7 Variation of dispersion with wavelength for **a** e = 0.4, 1.2, 2.0, 2.8, and 3.6 mm, **b** a = 2, 3, 4, 5, and 6 mm **c** d = 1.4, 1.8, 2.2, 2.6, and 3.0 mm

at 3 and 4 mm wavelength, respectively. In Fig. 7c, we have plotted the evaluation of the dispersion curve for various values of 'd'. As we increase 'd' values from 1.4 to 3 mm, we observed a very small change in the dispersion curve. The dispersion is directly proportional to the effective index of the mode. So, minute changes in the value of the effective index for changes in 'd' values result in small changes in dispersion values. Effective index at 3, 4, and 6 mm at d = 1.4–2.0 mm has a difference of just 0.0022, 0.0035, and 0.0065 respectively. Whereas, the difference in effective index with change in 'e' value at mentioned wavelengths is 0.019, 0.292, and 0.0488, respectively.

Next, we have calculated the tolerance study of loss for various waveguide parameters. Analysis of the loss curve for different values of 'e' shows very important characteristics of the present waveguide. We have plotted the loss curve for various values of 'e', i.e., 0.4, 1.2, 2.0, 2.8, and 3.6 mm and as shown in Fig. 8a. For e = 3.6 mm, the curve shows a 1.1 dB/cm loss at 3 mm wavelength, and 1.52 dB/cm loss at 4 mm wavelength. As we decrease the 'e' value from 3.6 to 0.4 mm, the loss component increases from 1.1 to 3.37 dB/cm at a wavelength of 3 mm. For e = 2 mm, the structure shows minimal values of dispersion and loss at 3 mm wavelength. In Fig. 8b, we have shown the effect of 'd' on the loss curve. Changing the value of 'd' from 1.4 to 3 mm, the design doesn't have a significant effect up to a wavelength region of 5.5 µm. Therefore, for d = 1.4 mm, the mode shows a loss of 1.9 dB/cm at 3 mm wavelength and a loss of 2.28 dB/cm at 4 mm wavelength. For d = 3 mm, the mode shows a loss of 1.82 dB/cm at 3 mm wavelength and a loss of 2.17 dB/cm at 4 mm wavelength. The effect of changing 'd' in small steps doesn't make difference in loss because a majority of the mode field is well confined in the middle and lower slabs. In Fig. 8c, we reported the loss as a function of wavelength for different values of 'a', i.e., 2, 3, 4, 5, and 6 mm. As we decrease 'a' from 6 to 2 mm, the loss value increases from 1.36 to 2.51 dB/cm at a wavelength of 3 mm. The proposed structure results in a high loss at 4 mm wavelength for different values of 'a'. Decreasing the value of 'a' increases the magnitude of the loss but the trend of the curve remains similar. The designed waveguide shows better results as compared to the previously reported designs [19].

From the above tolerance studies, we can conclude that with the optimized waveguide parameters (a = 4 µm, b = 8 µm, c = 15 µm, d = 2.2 µm, e = 2 µm, f = 1.5 µm, g = 9 µm, and h = 5.55 µm), we can achieve low dispersion and low loss at 3 mm wavelength.

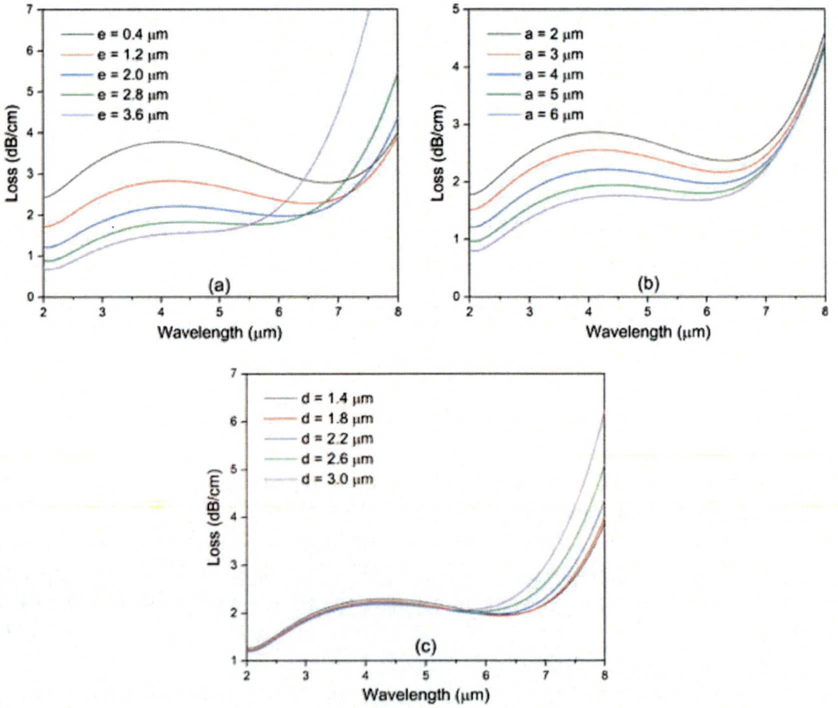

Fig. 8 Variation of loss with wavelength for **a** e = 0.4, 1.2, 2.0, 2.8, and 3.6 mm, **b** a = 2, 3, 4, 5, and 6 mm **c** d = 1.4, 1.8, 2.2, 2.6, and 3.0 mm

4 Conclusions

In the present paper, the proposed structure for low dispersion and low loss was designed, demonstrated, and studied numerically. The structure was designed by considering $Ge_{11.5}As_{24}Se_{64.5}$ as the core material and it consists of three rectangular slabs placed on top of one other. Cladding is designed with $Ge_{11.5}As_{24}S_{64.5}$ material and it is placed around the core material. We numerically evaluated the effective index and dispersion of the mode. The present design achieves a low dispersion value of − 37.9 ps/nm km at a wavelength of 3 μm, along with the structure showing a low loss value of 1.84 dB/cm at the same wavelength. Our simulation results revealed the low loss for the fundamental mode at the 3 μm region. For the proposed structure at 4 μm wavelength, we have also investigated the dispersion and loss properties. Designed waveguide results in a high loss around 4 mm wavelength region. Therefore, the present waveguide configuration offers great flexibility in manipulating the loss of the mode and dispersion. Numerical results presented in the paper provide significant importance in order to achieve low loss value and good dispersion value at 3 μm wavelength. Such a waveguide structure should find applications in supercontinuum generation sources.

Acknowledgements We would like to thank the Vellore Institute of Technology, Vellore, for their assistance with the software. We wish to extend our special thanks to VIT-AP University for its financial support.

References

1. Kogelnik H (1988) Theory of optical waveguides. Springer-Verlag, Germany, Berlin, pp 7–88
2. Tien PK (1977) Integrated optics and new wave phenomena in optical waveguides. Rev Mod Phys 49:361–420
3. Calvo ML, Lakshminarayanan V (2007) Optical waveguides: from theory to applied technologies. CRC Press, Boca Raton
4. Selvaraja SK, Sethi P (2018) Review on optical waveguides. In: Emerging waveguide technology. Intech Open, London, UK, p 95
5. Atakaramians S, Afshar V, Monro TM, Abbott D (2013) Terahertz dielectric waveguides. Adv Opt Photonics 5:169–215
6. Chen F, Wang XL, Wang KM (2007) Development of ion-implanted optical waveguides in optical materials: a review. Opt Mater 29:1523–1542
7. Jayakrishnan K, Hitaishi V, Ashok N (2022) Slot waveguide microring resonator based on silicon nitride for refractive index sensing. In: 2022 IEEE international conference on nanoelectronics, nanophotonics, nanomaterials, nanobioscience and nanotechnology (5NANO). IEEE, pp 1–3
8. Chan WK, Yi-Yan A, Gmitter TJ, Florez LT, Jackel JL, Yablonovitch E, Bhat R, Harbison JP (1990) Optical coupling of GaAs photodetectors integrated with lithium niobate waveguides. IEEE Photonics Technol Lett 2:194–196
9. Bogaerts W, de Heyn P, Vaerenbergh TV, de Vos K, Selvaraja SK, Claes T, Dumon P, Bienstman P, Van Thourhout D, Baets R (2012) Silicon microring resonators. Laser Photonics Rev 6:47–73

10. Uddin MA, Maswood MMS, Dey UK, Alharbi AG, Akter M (2020) A novel optical micro ring resonator biosensor design using lithium niobate on insulator (LNOI) to detect the concentration of glucose. In: 2nd novel intelligent and leading emerging sciences conference, NILES 2020. Institute of Electrical and Electronics Engineers Inc., pp 350–354
11. Bogaerts W, Selvaraja SK, Dumon P, Brouckaert J, de Vos K, Van Thourhout D, Baets R (2010) Silicon-on-insulator spectral filters fabricated with CMOS technology. IEEE J Sel Top Quantum Electron 16:33–44
12. Neethish MM, Acharyya JN, Kiran PP, Prakash GV, Sharan A, Kumar VVRK (2022) Broad white light supercontinuum generation in Barium Zinc Borate glasses. J Lumin 251:119190
13. Zhang J, Jiang J, Wang K, Ishihara H, Shimada K, Umeda S, Yokoyama N, Honda H, Kurose K, Kawata Y, Sugita A, Inoue Y, Uemukai M, Tanikawa T, Katayama R, Nakano T (2022) Fabrication and evaluation of rib-waveguide-type wavelength conversion devices using GaN-QPM crystals. Jpn J Appl Phys 61:SK1020
14. Eisenberg HS, Morandotti R, Silberberg Y, Arnold JM, Pennelli G, Aitchison JS (2002) Optical discrete solitons in waveguide arrays. I. Soliton formation. J Opt Soc Am B 19:2938
15. Fukuda H, Yamada K, Shoji T, Takahashi M, Tsuchizawa T, Watanabe T, Takahashi JI, Itabashi SI, Lee DKKR, Lim HC, Luan A, Agarwal J, Foresi LC (2005) Four-wave mixing in silicon wire waveguides. Opt Express 13:4629–4637
16. Frigg A, Boes A, Ren G, Nguyen TG, Choi DY, Gees S, Moss D, Mitchell A (2020) Optical frequency comb generation with low temperature reactive sputtered silicon nitride waveguides. APL Photonics 5(011302):1–6
17. Serna S, Lin H, Alonso-Ramos C, Yadav A, le Roux X, Richardson K, Cassan E, Dubreuil N, Hu J, Vivien L (2018) Nonlinear optical properties of integrated GeSbS chalcogenide waveguides. Photonics Res 6:B37–B42
18. Zhai Y, Yuan C, Qi R, Zhang W, Huang Y (2015) Reverse ridge/slot chalcogenide glass waveguide with ultrabroadband flat and low dispersion. IEEE Photonics J 7:7801609
19. Ashok N, Lee YL, Shin WJ (2017) Chalcogenide waveguide structure for dispersion in mid-infrared wavelength. Jpn J Appl Phys 56(032501):1–5
20. Karim MR, Ahmad H, Ghosh S, Rahman BMA (2018) Design of dispersion engineered As_2Se_3 channel waveguide for mid-infrared region supercontinuum generation. J Appl Phys 123:213101
21. Xia D, Huang Y, Zhang B, Yang Z, Zeng P, Shang H, Cheng H, Liu L, Zhang M, Zhu Y, Li Z (2021) On-chip broadband mid-infrared supercontinuum generation based on highly nonlinear chalcogenide glass waveguides. Front Phys 9:598091
22. Zhang L, Yue Y, Li YX, Beausoleil RG, Willner AE (2009) Highly dispersive slot waveguides. Opt Express 17:7095–7101
23. Zhu M, Liu H, Li X, Huang N, Sun Q, Wen J, Wang Z (2012) Ultrabroadband flat dispersion tailoring of dual-slot silicon waveguides. Opt Express 20:15899
24. Mann V, Ashok N, Rastogi V (2015) Coupled strip-slot waveguide design for dispersion compensation. Opt Quantum Electron 47:3161–3169
25. Huang J, Ye F, Li Q (2022) Simultaneous pulse combination and nearly self-similar pulse compression in tapered silicon waveguides at around 2.0 μm. IEEE Photonics J 14(4):1–7
26. Liu M, Gu C, Fan X, Li Z, Huang H, Lu Z, Zhao W (2022) Efficient dispersion engineering for three-octave-spanning supercontinuum generation in nanophotonic waveguides. Opt Laser Technol 150:107923
27. Ma P, Choi DY, Yu Y, Gai X, Yang Z, Debbarma S, Madden S, Davies BL (2013) Low-loss chalcogenide waveguides for chemical sensing in the mid-infrared. Opt Express 21:29927–29937
28. Yu Y, Gai X, Ma P, Vu K, Yang Z, Wang R, Choi DY, Madden S, Davies BL (2016) Experimental demonstration of linearly polarized 2–10 μm supercontinuum generation in a chalcogenide rib waveguide. Opt Lett 41:958–961
29. Gai X, Han T, Prasad A, Madden S, Choi DY, Wang R, Bulla D, Davies BL (2010) Progress in optical waveguides fabricated from chalcogenide glasses. Opt Express 18(25):26635–26646

Detection of Pathological Conditions in Nail Samples Using Laser-Induced Breakdown Spectroscopy

K. Rithika, R. Sowmya, G. Rithick kumar, M. Thangaraja, Pauline John, and V. Sathiesh Kumar

Abstract In this paper, laser-induced breakdown spectroscopy (LIBS) technique is used to analyze human fingernails using nanosecond laser pulses from an neodymium-doped yttrium aluminum garnet laser (Nd:YAG) at 1064 nm. Nail as a biosample has various advantages when compared to other biological samples. Ca, Mg, Fe, and other elements found in the body can also be found in the nail, but in varying concentration. The emission spectrum from nail samples was collected in the spectral range of 180–900 nm, and it revealed the presence of elements like Mg, Zn, Fe, Mn, P, I, Na, Ca, K. Nail samples were collected from different subjects based on age, gender, and medical conditions. A total of 27 samples were collected; 17 of them were collected from normal people, and 10 of them were collected from people with different pathological conditions like diabetes, thyroid, PCOS, wheezing, etc. The results obtained in abnormal samples were analyzed and verified.

Keywords Laser-induced breakdown spectroscopy · Nail samples · Elemental analysis · Pathological conditions

K. Rithika (✉) · R. Sowmya · G. R. kumar · P. John
Department of Biomedical Engineering, Sri Sivasubramaniya Nadar College of Engineering, Chennai, India
e-mail: rithika1943@bme.ssn.edu.in

R. Sowmya
e-mail: sowmya1957@bme.ssn.edu.in

G. R. kumar
e-mail: rithick1942@bme.ssn.edu.in

P. John
e-mail: paulinejhn@gmail.com

M. Thangaraja · V. S. Kumar
Department of Electronics Engineering, Madras Institute of Technology, Chennai, India
e-mail: rajaeee83@yahoo.co.in

V. S. Kumar
e-mail: sathiieesh@gmail.com

N. M. Rao et al. (eds.), *Advanced Nanomaterials and Their Applications*,
Springer Proceedings in Materials 22, https://doi.org/10.1007/978-981-99-1616-0_10

1 Introduction

An optical technique based laser-induced breakdown spectroscopy (LIBS) employs the phenomena of optical emission after pulsed-laser ablation of a sample surface to pinpoint and examine specific elements in the sample. It is possible to scan solid surfaces using laser pulses while maintaining spatial information [1, 2]. It does not necessitate any specific sample pretreatment [1–3]. In order to achieve high spatial resolution on the target, a highly focused laser beam is directed onto the sample's surface. Finally, the analysis detects the vast majority of elements without bias [1–3]. Previously, LIBS was used to analyze the elements in biological samples [1, 4]. Laser-induced breakdown spectroscopy has been proposed for the analysis of teeth, hair, and nails. An individual's nails contain biological data about them. Among other things, nails store information about nutrition, habitat, and the environment [1–4]. There are several things that nails store, including nutrition, habitat, and environment. As a result of a variety of mechanisms, including protein synthesis and chemical bonding, human nails are composed primarily of keratin-rich proteins that incorporate trace elements based on dietary intakes and other exposures [4, 5]. Consequently, clinical studies have recently begun to pay more attention to finger and toe nails, which are particularly useful trace element markers [5–7]. The element composition of nails is less likely to change, and the chemicals added by nail polishing may be removed in a lab. The collection of nail samples is also simple and does not require special storage facilities. Elements found in the body (e.g., Ca, Mg, Fe, and so on) can also be found in the nail, but at varying levels [4–8]. Several nail samples are collected randomly with the consent of the subjects and discarded after use. The data obtained are processed to determine whether there is a discriminating factor between LIBS spectra of nails and the subject's sex and age. Until now, the possible links between nail minerals and thyroid disease have received little attention. In this study, elements in fingernails from healthy and thyroid disease subjects were analyzed using laser-induced breakdown spectroscopy. In order to achieve efficient results, it is beneficial to classify nails according to their age and gender. According to the research, males had higher magnesium concentrations than females. Both genders show a decrease in Ca, Fe, Zn, Mg, and K elements with age. LIBS is an effective method for analyzing nails and identifying nail disorders. Furthermore, LIBS can be used to assess conditions on other planets, evaluate artifact restoration quality, control the quality of building and glass restoration, analyze human biological samples, and examine deep-sea objects remotely [9]. Laser-induced breakdown spectroscopy offers many advantages over the next line should be continuous.

X-ray photoelectron spectroscopy (XPS) in terms of ease of use, sample preparation, and real-time results [10]. Samples collected for the study, experiment setup, methodology and results are presented in the following sections.

Fig. 1 Hydraulic press

2 Methodology

2.1 Samples

During the study, nails were collected from 27 individuals of different ages and genders, which included 17 normal samples and 10 abnormal samples with different pathological conditions like diabetes, thyroid, PCOS, wheezing, etc. The individual nail samples were kept in non-reactive, self-sealing plastic envelopes at room temperature. The curved surface of nail samples was flattened using hydraulic press as shown in Fig. 1. The flattened nail samples were further used for LIBS studies.

2.2 Experimental Setup and Methodology

Figure 2 shows the LIBS setup used in this experiment. It consists of a Nd:YAG laser source (Q-Smart-450 mJ) with a wavelength of 1064 nm, maximum energy per pulse of 450 mJ, and a pulse repetition rate of 10 pulses per second. The beam was focused by a plano-convex quartz lens with a focal length of 300 mm, on the nail sample (formed into pellets using hydraulic press), mounted on a rotating mount. Using optical fiber, light emitted from the plasma was captured and fed into a spectrometer (Ocean FX). A personal computer (PC) was used to analyze the emission spectrum captured by the spectrometer (Ocean FX). Using the PC, data were acquired and

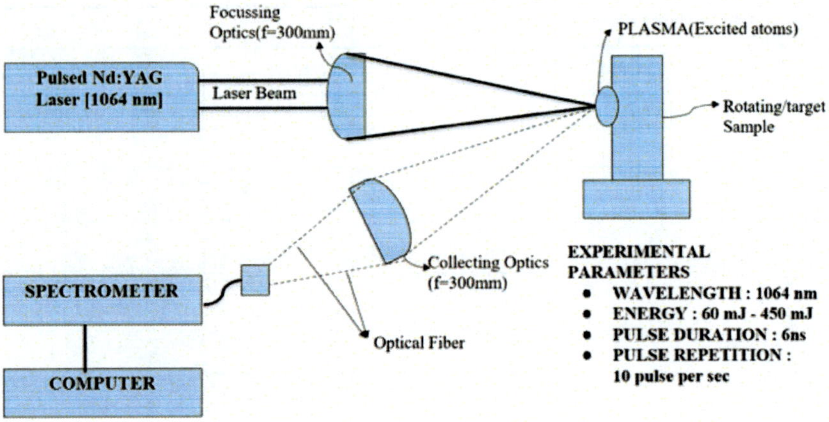

Fig. 2 Experimental setup of LIBS

analyzed. The laser beam was directed at the nail sample to ablate it and create plasma. A specific amount of mass is not required; the flattened nail was ablated at different locations, and the averaged spectra was used.

3 Results and Discussion

The results obtained from the spectrometer for various nail samples were collected and plotted. Figure 3 shows a typical elemental information obtained for a normal human nail, for a wavelength range of 100–1000 nm. Prominent peaks obtained were compared with the NIST database [11].

Comparison was done for two diabetic and two thyroid samples that were collected. Age-wise and gender-wise comparison is also performed.

A correlation in the elemental peaks obtained from a normal nail sample as shown in Fig. 3 with that of the data obtained from the NIST database as shown in Table 1 was achieved [11].

3.1 Age-Wise Comparison

Samples were collected from different age groups in the range of 19–80 years. The intensity levels of elements were compared between the different age groups.

It is evident that the elemental composition is the same between the samples of the same age (sample age 21) as shown in Fig. 4, with no distinct differences found in the peaks. It can also be seen that there are distinct differences in peaks for various elements in different age groups as shown in normalized spectra of Fig. 5. The blue

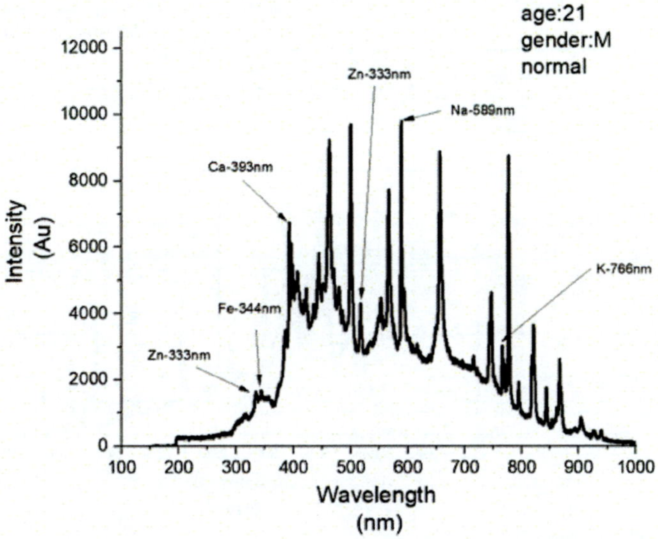

Fig. 3 LIBS spectra from nail sample of normal subject

Table 1 Elements and their corresponding wavelengths obtained from NIST database [11]

Element	Wavelength (nm)
Zinc	333
Ferrous	344–358
Calcium	393.6–399
Potassium	766–769
Sodium	589

trace in Fig. 5 corresponds to the nail sample with age 47, and black trace corresponds to nail sample with age 21. Elements like Fe, Zn, Ca, Mg are predominantly evident in the results that indicate deficiency in a particular element in the elderly due to various conditions like poor absorption, malnutrition, medications, etc.

To know about the relationship between age and element deficiency levels, a regression graph was plotted using the Mg levels of different age groups as shown in Fig. 6.

X indicates the age of different samples, and Y represents the intensity levels of magnesium. The regression graph was plotted using MINITAB. Positive correlation and strong significance was obtained showing that the intensity levels and age can be compared and are dependent on each other. Based on the regression as shown in Fig. 6, we can see strong significance with age and intensity levels of the elements. Figure 7 shows the relative intensity comparison of the deficit elements like Zn, Mg, Fe, Ca with respect to Na, and it is higher in case of normal sample and lower in aged sample.

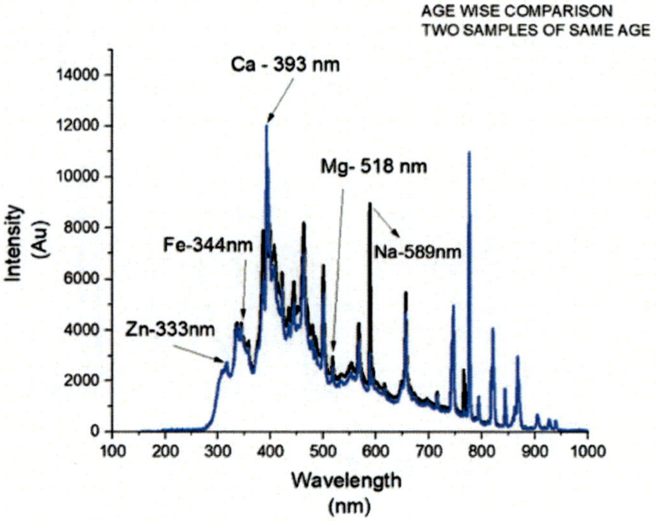

Fig. 4 Same age analysis (age 21)

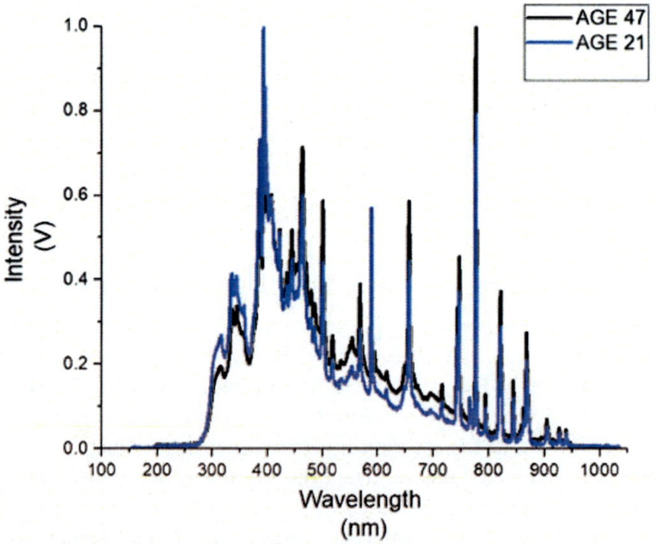

Fig. 5 Different age analysis

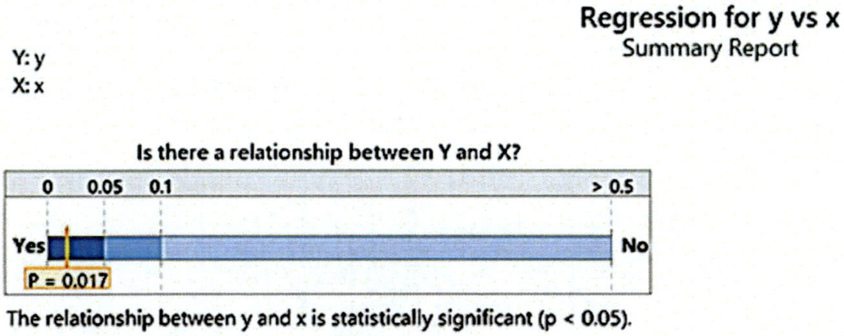

Fig. 6 Regression for age-wise comparison (MINITAB)

Fig. 7 Scatter plot

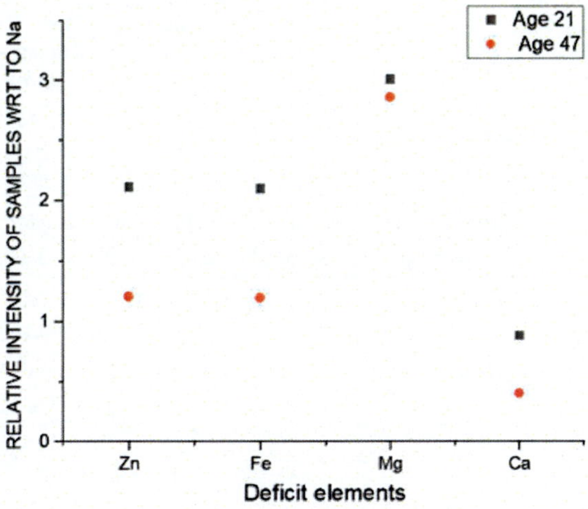

3.2 Gender-Wise Comparison

Samples collected are segregated based on gender, and their results were analyzed. Similar changes in peaks associated with Ca, Zn, Fe, and Mg as shown in Fig. 8 were observed as noticed in age-wise comparison, where the blue trace in the graph shows spectrum of a female sample and black trace graph shows spectrum of male sample.

Mg intensity levels were higher in males when compared to females of different age groups as shown in Table 2. Similar trends were found in other elements too [12].

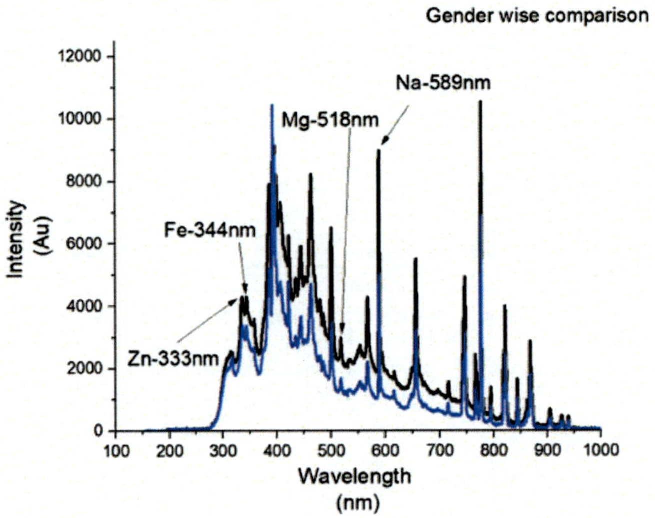

Fig. 8 Gender-wise analysis

Table 2 Intensity of Mg obtained in different samples

Intensity spectrum obtained in females	Age	Intensity spectrum obtained in males	Age
1272	19	4573	21
2976	21	4262	21
2602	21	4654	24
2531	21	5052	35
5138	36	5585	47

3.3 *Thyroid Analysis*

It was observed that changes in levels Na, K, I can indicate presence of hyper- or hypothyroidism [13, 14]. This can be observed in the peaks of the nail samples of subjects with thyroid condition, when analyzed for potassium and iodine as shown in Figs. 9 and 10, respectively. The red, blue, black traces in the graph indicate 3 trials carried out using the nail samples of subjects with thyroid condition.

Table 3 shows the comparison of intensity levels of iodine and potassium in normal and thyroid samples. Similar trends in peaks were obtained for sodium and calcium when comparing nail samples of a normal subject with a subject with thyroid condition, as shown in Figs. 11 and 12.

Sodium, potassium, iodine and calcium are important factors of the thyroid gland [14]. In other cases, low potassium and sodium impairs thyroid hormone activity at the cellular level [14]. The production of T3 and T4 requires iodine [15]. Inappropriate

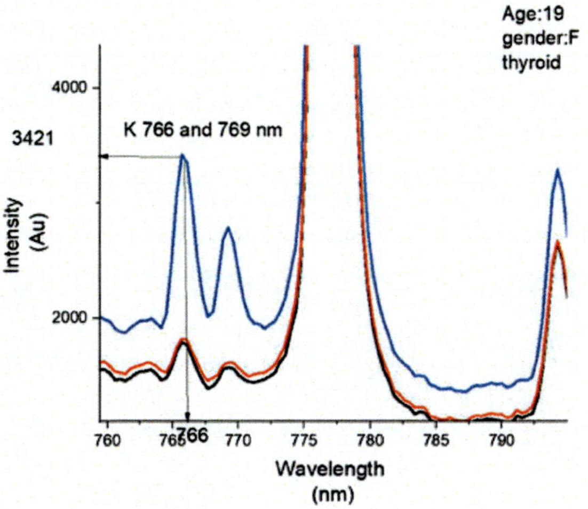

Fig. 9 Potassium analysis in nail sample

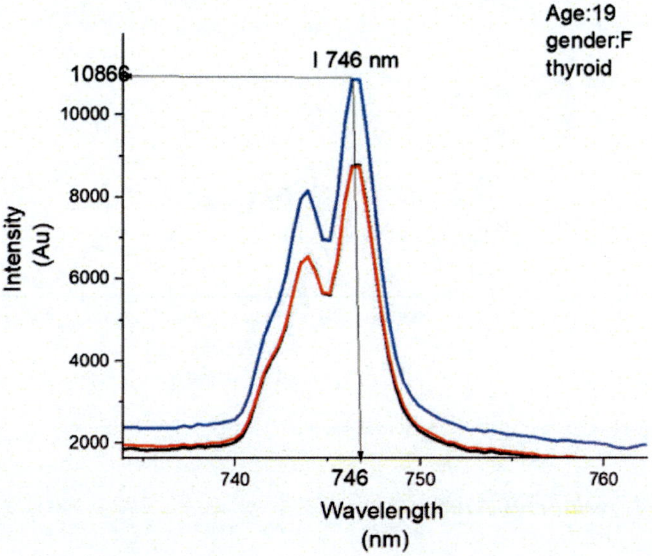

Fig. 10 Iodine analysis in nail sample

levels of these factors can cause many thyroid disorders, including hypothyroidism and hyperthyroidism. The intensities of Ca, Na, I, K are significantly higher in case of a thyroid sample hence indicating the presence of the disease.

Table 3 Potassium and iodine intensity comparison in normal and abnormal samples

	Wavelength (nm)	Intensity in normal sample (Au)	Intensity in abnormal sample (Au)
Potassium analysis			
Trial 1	766	1785	3106
Trial 2	766	1215	1725
Trial 3	766	915	1544
Iodine analysis			
Trial 1	746	5005	10,866
Trial 2	746	4966	8769
Trial 3	746	4836	8751

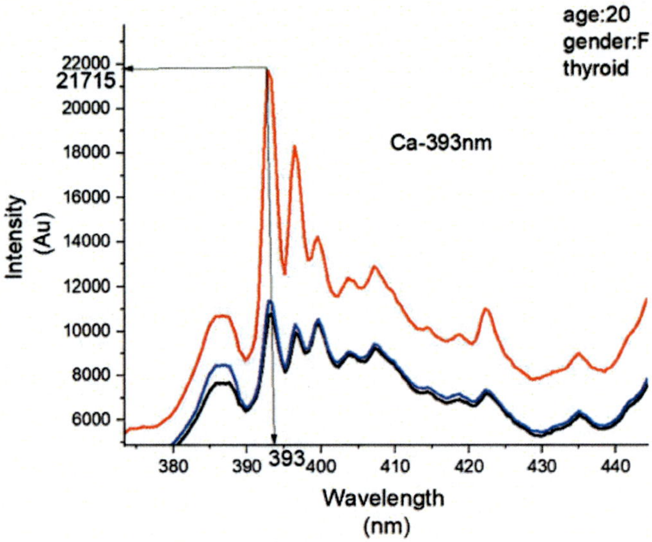

Fig. 11 Calcium analysis in thyroid sample

3.4 Diabetes Analysis

Magnesium is one of the important biomarkers in predicting diabetes [15], and hence, the Mg levels were analyzed as shown in Figs. 13 and 15, and the Mg intensity of each sample specifically at 517 nm is plotted in Figs. 14 and 16. Figures 13, 14, 15, and 16 are normalized results.

When comparing the intensity values of normal and diabetic samples, it is clear that the deficiency of Mg is present in the diabetic sample. Magnesium is one of the most prevalent ions in living cells, and in healthy individuals, its blood plasma

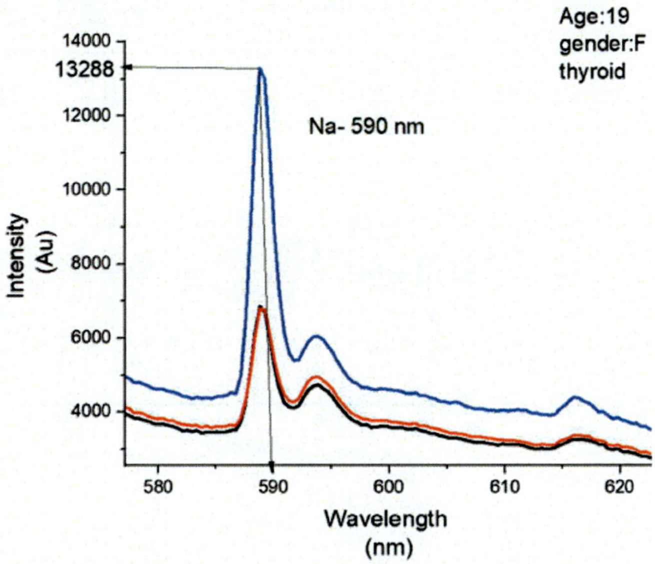

Fig. 12 Sodium analysis in thyroid sample

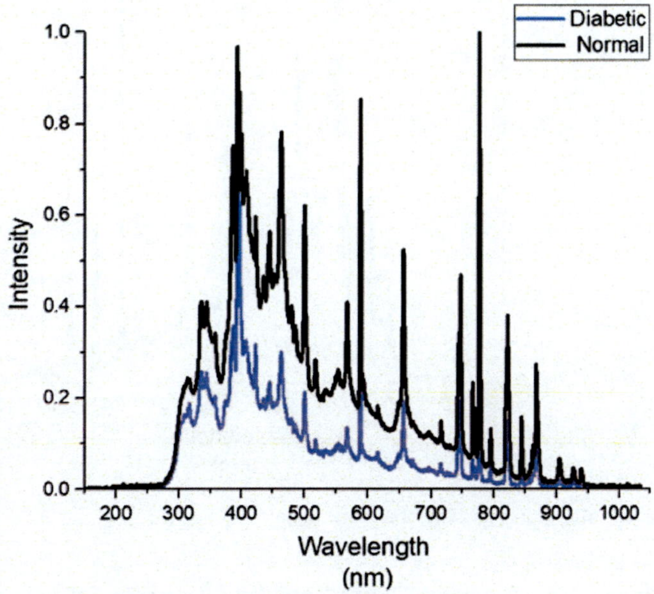

Fig. 13 Magnesium analysis in diabetic sample 1

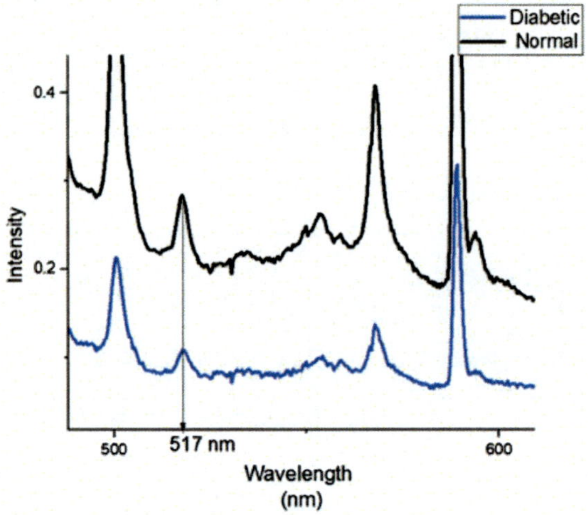

Fig. 14 Peak variation in intensity of Mg for sample 1

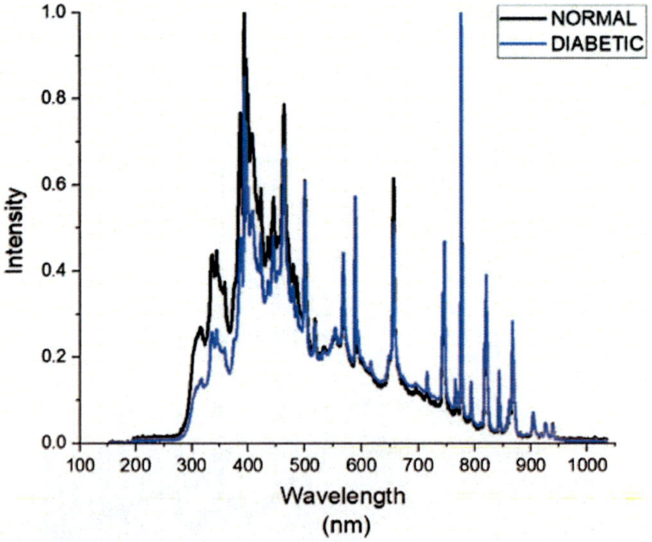

Fig. 15 Magnesium analysis of diabetic sample 2

concentration is constant. Diabetes problems have been linked to magnesium deficiency [16]. The findings show that the Mg level in nails was less in the diabetes group. Hence, the analysis of Mg could help in detecting the presence of diabetes as shown in Table 4.

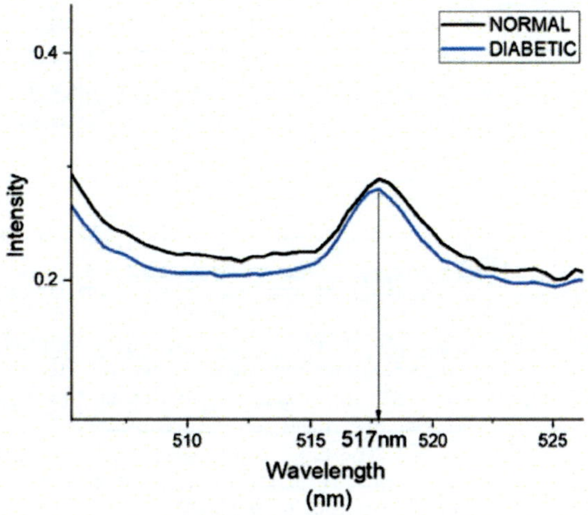

Fig. 16 Peak variation in intensity of Mg for sample 2

Table 4 Intensity comparison of magnesium in normal and abnormal sample

	Wavelength (nm)	Intensity in normal sample (Au)	Intensity in diabetic sample (Au)
Trial 1	517.8	4049	2474
Trial 2	517.8	4104	2848
Trial 3	517.8	4296	3240

4 Conclusion and Future Scope

LIBS was performed on various nail samples collected based on age, gender, and pathological conditions. The samples were segregated based on normal and abnormal conditions. Analysis was done for detection of pathological conditions. Graphs were plotted based on the data collected from experiments using Origin software. Comparing the results of normal and abnormal samples in diabetes and thyroid, significant trends were seen in the intensity values of the corresponding elements. Comparison was done based on age, and regression graph was plotted, where in the R-squared value was found to be high and the p value showed a strong significance, and the correlation coefficient being positive helps conclude that both are correlated and can be used to determine the trends in elemental composition and the corresponding age-related factors. Similarly, significant trends were seen when comparison based on gender was performed. Age and gender can also be factors of a particular disease which can also be seen in the graphs plotted for various elements. Hence, LIBS on nail samples helps to detect the pathological conditions with high specificity. In future, analysis can be done with a larger set of nail samples. A more accurate comparison can be done with a large dataset from a large number of samples using statistics. Analysis of nail samples including alcoholics, chain-smokers, kidney patients, tobacco addicts, cancer patients, etc., can be performed in future. Further, collection of a large set of samples for more diverse statistical analysis could be performed.

References

1. Harun HA, Zainal R, Daud YM (2017) Analysing human nails composition by using laser induced breakdown spectroscopy. Sains Malaysiana 46(1):75–82
2. Fortes FJ, Moros J, Lucena P, Cabalín LM, Laserna JJ (2013) Laser-induced breakdown spectroscopy. Anal Chem 85(2):640–669
3. Kaiser J, Novotný K, Martin MZ, Hrdlička A, Malina R, Hartl M, Adam V, Kizek R (2012) Trace elemental analysis by laser-induced breakdown spectroscopy—biological applications. Surf Sci Rep 67(11–12):233–243
4. He K (2011) Trace elements in nails as biomarkers in clinical research. Eur J Clin Invest 41(1):98–102
5. Nasli-Esfahani E, Faridbod F, Larijani B, Ganjali MR, Norouzi P (2011) Trace element analysis of hair, nail, serum and urine of diabetes mellitus patients by inductively coupled plasma atomic emission spectroscopy. J Diabetes Metab Disord 10:5
6. Zhang S, Hu Z, Zhao Z, Chen F, Tang Y, Sheng Z, Guo L (2021) Quantitative analysis of mineral elements in hair and nails using calibration-free laser-induced breakdown spectroscopy. Optik 242:167067
7. Rusak DA, Zeleniak AE, Obuhosky JL, Holdren SM, Noldy CA (2013) Quantitative determination of calcium, magnesium, and zinc in fingernails by laser-induced breakdown spectroscopy. Talanta 117:55–59
8. Pathak AK, Kumar R, Singh VK, Agrawal R, Rai S, Rai AK (2012) Assessment of LIBS for spectrochemical analysis: a review. Appl Spectrosc Rev 47(1):14–40
9. Hark RR, East LJ (2014) Forensic applications of LIBS. In: Laser-induced breakdown spectroscopy. Springer, Berlin, Heidelberg, pp 377–420
10. Ledesma R, et al (2019) Correlation of trace silicone contamination analyses on epoxy composites using X-ray Photoelectron Spectroscopy (XPS) and Laser-Induced Breakdown Spectroscopy (LIBS). Appl Spectrosc 73(2):229–235
11. NIST database link. https://physics.nist.gov/PhysRefData/Handbook/Tables/findinglist.htm
12. Chaudhary K, Ehmann W, Rengan K, Markesbery W (1995) Trace element correlations with age and sex in human fingernails. J Radioanal Nucl Chem 195(1):51–56
13. Ahmed I, Ahmed R, Yang J, Law AWL, Zhang Y, Lau C (2017) Elemental analysis of the thyroid by laser induced breakdown spectroscopy. Biomed Opt Express 8:4865–4871
14. Bahreinian M, Tavassoli SH. Possibility of thyroidism diagnosis by laser induced breakdown spectroscopy of human fingernail
15. Burman KD, Wartofsky L (2000) Iodine effects on the thyroid gland: biochemical and clinical aspects. Rev Endocr Metab Disord 1(1–2):19
16. Bahreini M, Ashrafkhani B, Tavassoli SH (2013) Discrimination of patients with diabetes mellitus and healthy subjects based on laser-induced breakdown spectroscopy of their fingernails. J Biomed Opt 18(10):107006

A Review of mRNA Vaccines with the Aid of Lipid Nanoparticles

Simran Saikia, Shreya Barman, S. Sudhimon, M. Mukesh Kumar, G. Shanmugasundaram, and J. Sudagar

Abstract This review article highlights the importance of messenger ribonucleic acid (mRNA) vaccines and how it has been developed to fight against various diseases such as, human immunodeficiency virus (HIV), rabies, cancer treatments, and coronavirus (Covid-19). During the past two years, covid-19 has become a worldwide pandemic, and the mRNA has played a major role in the manufacturing of its vaccine. We have highlighted the technology behind the development of mRNA vaccine, synthesis, and working of the lipid nanoparticles (LNPs). This mRNA vaccine produces a duplicate of a molecule that corresponds to a viral protein for producing an immune response, and these are given to us in a series of shots designed to protect us from developing a disease. The LNPs which carry the mRNA protein prevent the degradation of it and maintain more constant serum levels. In addition, this review article specifically mentions HIV, rabies, cancer, covid-19 and how these are important in the treatment of these diseases. This review article further highlights the mRNA vaccines for the survival of human beings against various deadly diseases in the near future.

Keywords mRNA vaccine · LNPs · HIV · Rabies · Cancer · Coronavirus

---

S. Saikia · S. Barman
School of Electronics Engineering, VIT-AP University, Vijayawada, Andhra Pradesh 522241, India

S. Sudhimon · M. M. Kumar · J. Sudagar (✉)
Department of Physics, School of Advanced Sciences, VIT-AP University, Vijayawada, Andhra Pradesh 522241, India
e-mail: sudagar.jothi@vitap.ac.in

G. Shanmugasundaram
Department of Computer Science and Engineering, Chennai Institute of Technology, Chennai, Tamil Nadu 600069, India

N. M. Rao et al. (eds.), *Advanced Nanomaterials and Their Applications*,
Springer Proceedings in Materials 22, https://doi.org/10.1007/978-981-99-1616-0_11

1 Introduction

In the year 2019, the global pandemic of covid-19 hits the world, and the initial case of this deadly disease was found in the Hubei province of China. This disease causes severe acute respiratory problems, fever, cough, and other symptoms [1]. Because of this outbreak, 6,518,749 deaths and 613,410,796 confirmed cases were reported over the globe till 28th September 2022 [2]. Many clinical trials of vaccine manufacturing were being done to fight this pandemic until the discovery of the messenger ribonucleic acid (mRNA) vaccine in the year 2021. A vaccine is a biological material that is being introduced to a human's immune system to produce immunity to a particular disease. There are several types of vaccines that are created by scientists depending on three criteria: the way the immune system responds to a particular infectious agent, where living beings need to be vaccinated against the infections caused, and this is the best way to produce the vaccine. In general, the existing types of vaccines include inactivated vaccines, live-attenuated vaccines, mRNA vaccines, subunit, recombinant, polysaccharide, conjugate vaccines, toxoid vaccines, and viral vector vaccines [3, 4]. Among all the vaccines, the mRNA vaccine stands out, when it comes to curing diseases because it holds the promise to reform the field of medicine, offers noble vaccine composition in a rapid manner, and is proven clinically safe [5]. In this review article, we review the development of the mRNA vaccine over the years to cure many infectious diseases with the help of lipid nanoparticles (LNPs). Inspired by the wide range of applications in the manufacturing of m-RNA vaccines are being used in the aid of lipid nanoparticles. Over ~27,000 journals/articles/scientific documents regarding 'mRNA vaccines' searched in the Scopus database between 2013 and 2023 (last 10 years).

The recent rising trends of LNP vaccines can be identified in the present times. It is used in the preparation of cosmetics, used in medical imaging, in nano-reactors, etc. During COVID-19 vaccination, LNPs effectively protect and transport mRNA to cells [6]. In this review article, we mainly emphasize and review on the development of mRNA vaccines using the lipid nanoparticles of various diseases like human immunodeficiency virus (HIV), rabies, cancer, and covid-19. This article deals with how mRNA vaccines play an important role in curing of infectious and genetic diseases where mRNA leads to the future foundation for the development of new medications.

2 Technology Behind the Development of mRNA Vaccine

The mRNA and liposomes were discovered in 1960, and it took years for the researchers to finally develop a vaccine from it [7]. The plasmid deoxyribonucleic acid (pDNA) preparation is the initial step for the production of the mRNA vaccine. To construct an mRNA vaccine, pDNA contains a deoxyribonucleic acid (DNA)-dependent RNA polymerase promoter, namely T7. In general, unpolished pDNA carries remains of bacterial genomic DNA, and its three types are relaxed circle, supercoiled, or linear in variable proportion. Therefore, we require a pure and invariant form of pDNA for the reproducible preparation of the vaccine. The next step is to linearize the pDNA, which acts like a prototype as the DNA-dependent RNA which polymerase in order to reproduce the mRNA and followed by the degradation of a deoxyribonuclease (DNase) process step. There are two types of mRNA constructs that are being constantly estimated which are non-replicating mRNA (NRM) constructs and self-amplifying mRNA (SAM), each of the construct have an open reading frame, a cap structure, a 3′ poly(A) tail and 5′ and 3′ untranslated regions (UTRs). In the manufacturing process of mRNA, the 5′ caps and the 3′ poly(A) tail are accomplished during the in vitro transcription pace or enzymatically following the transcription (see Fig. 1). The incorporation of the enzyme can be achieved by using 2′-O-methyltransferase, and guanylyl transferase gives a Cap 1 or Cap 0 structure, correspondingly and the poly-A tail can be attained via enzymatic addition through poly-A polymerase [5, 8].

Purification is one of the pivotal steps in the manufacturing of the mRNA. During this process, raw materials are removed which are used for the amalgamation, along with the mRNA impurities such as double-stranded RNA, aggregates, and aborted (short) transcripts by the technique of high-pressure liquid chromatography (HPLC). In this process, the produced drug substance is transferred into a drug product and then released based on some factors such as their identity, purity, potency testing, and sterility. These manufacturing processes have an upper hand such that it is easy to transfer into a new vaccine within a short span of time [8].

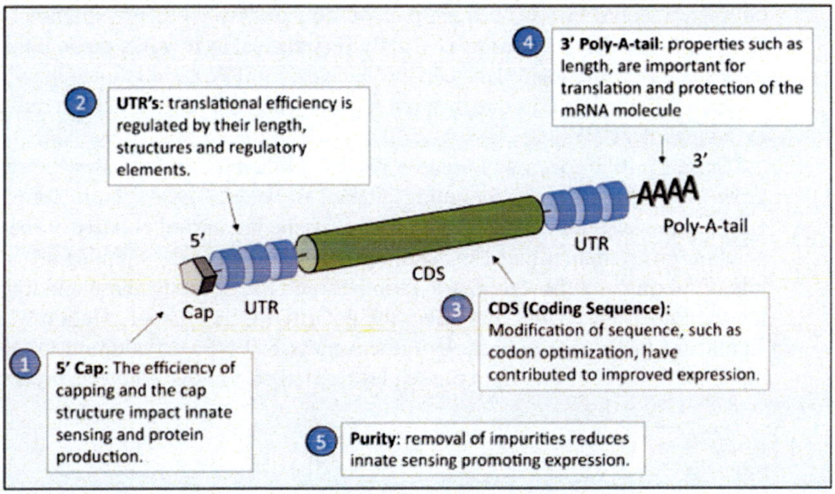

Fig. 1 Schematic mRNA construct [8] [Adapted from Nicholas AC Jackson et al. (2020) (CC BY 4.0)]

The mRNA vaccine is designed to boost our immune system, but they act in a unique way compared to other vaccines. Due to its uniqueness, the mRNA vaccine doesn't contain any virus; instead they contain messenger molecules that give our body instructions to efficiently act against a disease. This instruction gives our cells an idea to make a piece of protein from the virus such as in the case of the virus that causes covid-19. This vaccine is designed in such a way that once it is delivered into the cytoplasm of a cell it productively utilizes the translational technology of the host cell. As a result, it generates an abundant amount of the predetermined immunogen that is given to the immune system in a proper manner [8, 9]. If you see the Table 1 (the history of mRNA vaccine), it clearly demonstrates that 'The real winner here is modified RNA.'

3 LNPs in mRNA Vaccine

3.1 Synthesis of LNPs

The LNPs are the nanoparticles in the mRNA vaccine which are also known as liposomes. Liposomes act as a carrier whose job is to deliver the viral protein in protective droplets form that target the effective area to trigger the production of antibodies, and it also has the capability to interact with the immune system. The LNPs synthesis process is carried out by condensing lipids that are amphiphilic molecules which are composed of three parts: a polar head group, a hydrophobic tail region, and a bridge between the two domains from ethanol solution [10, 11]. This method of LNP synthesis is usually very critical as it works on both the size of LNP and encapsulation efficacy. In this technique, mRNA is dissolved in an aqueous solution followed by the condensation process where it is encapsulated or complexes to the finished LNPs [12].

These LNPs are formed based on disk-like bilayered fragments which fuse and grow to bigger rafts when diluting ethanol in water. The structure forms closed LNPs at low ethanol concentration. However, the increasing polarity of the ethanol solution gives smaller particles, and as a result, it gives less significant LNPs. The rate of mixing and the volumetric ratio between the aqueous phase and lipid phase are the important factors that affect the polarity of the ethanol solution which is a significant factor that defines the effectiveness of LNPs. In addition to water, the quick mixing of ethanol-lipid phase plays a vital role in the synthesis process [10].

Table 1 History of mRNA vaccine [7]

Approximate year	mRNA	Lipid-based delivery system	mRNA with lipids (for vaccines)	Important remarks
~1960	mRNA discovered			
~1965		First liposomes (fatty bubbles, composed of lipid molecules) produced		
~1970	First proteins produced from isolated mRNA in lab	Liposomes used for drug delivery		
~1975		Liposomes used for vaccine delivery	First liposome-wrapped mRNA delivery to cells	
~1980	mRNA synthesized in lab			
~1985			Synthetic mRNA in cationic liposomes (structures made of positively-charged lipids) delivered to human cells, frog embryos	
~1990	mRNA tested as a treatment (in rats)		Liposome-wrapped mRNA delivered to mice	
~1995	mRNA tested as cancer vaccine (in mice)		First mRNA vaccines tested (for influenza, in mice)	Selected commercial R&D
~2000		First report of four-component lipid nanoparticles (at that time, to deliver DNA)		First mRNA-focused company founded (Merix Biosciences, later known as Argos, then CoImmune) + CureVac founded

(continued)

Table 1 (continued)

Approximate year	mRNA	Lipid-based delivery system	mRNA with lipids (for vaccines)	Important remarks
~2005	Discovery that modified RNA evades immune detection	Scalable method for manufacturing lipid nanoparticles		BioTech founded. Novartis and Shire establish mRNA divisions
~2010	First clinical trial of mRNA vaccine for infectious disease (rabies)		First mRNA vaccine in lipid nanoparticles tested in mice	Moderna founded
~2015		First drug with lipid nanoparticles (patisiran) approved	First clinical trial of mRNA vaccine in lipid nanoparticles (influenza)	US Defense Advanced Research Projects Agency begins funding mRNA vaccine research
~2020			mRNA-based COVID-19 vaccines win emergency authorization	

A long chain of scientific advances led to the first messenger RNA (mRNA) vaccines, released in 2020 to protect people against COVID-19. These vaccines, as well as mRNA drugs, make use of developments in the science of mRNA and in delivery systems, which are made of lipid molecules

3.2 Working of LNPs

The arrival of LNPs carrying mRNA has allowed a highly efficient new vaccine platform. Generally, there are four pillars in vaccine development: antigens, delivery-nanoparticles, vectors, and manufacturing. It has been found that a bit of RNA code is quite at risk since our body is instructed to demolish the free-floating RNA, so it requires a carrier to help them get directly into the cells. This problem is being solved while introducing a LNP which is perfect for encapsulating in order to turn into the delivery pillar and having massive expandable potential for manufacturing. LNPs could take a range of forms which enhances their capacity to compress an assortment of cargoes like peptides, genetic payloads like siRNA, saRNA, mRNA, and other small-scale molecules. The uttermost unique among them are those composed with ionizable cationic lipids. The LNP structure of the mRNA vaccines makes the way for the nanoprecipitation process for its creation. Precisely, nano-systems engage a microfluidic architecture that permits two streams to mix under laminar flow conditions enabling the process to be scalable and reproducible. The cationic lipids in an mRNA adjoin with the negative charges of the phosphate backbone of the mRNA [13]. The zwitterionic lipids help to stabilize the lipid bilayer of the LNP. Thus, it amplifies the transporting efficacy, and due to its aqueous environment, polyethylene glycol (PEG)-lipid forms the outer shell indicating its hydrophobic tails inward. It

Table 2 Advantages of mRNA vaccine over traditional vaccine [15, 16]

S. No.	mRNA vaccine	Traditional vaccine
1	mRNA vaccine comprises micro-ribonucleic acid (blueprint of protein)	Traditional vaccines consists of microbial protein or inactive microbes
2	Vaccine production is faster	Vaccine production is comparatively slower
3	mRNA molecules are simpler to produce	Hard to produce the right type of protein
4	Non-infectious and having no concern in DNA integration	Vaccines are infectious
5	mRNA vaccines are more stable as it increases the cell delivery efficiency	Traditional vaccines are comparatively less stable

reduces specific absorption of plasma proteins and forms a hydration cover over the nanoparticle which improves the colloidal constancy in the biological environment.

Finally, it creates unilamellar, predictably sized, uniform LNPs in just a few seconds. The LNPs always have an upper hand in the development of mRNA vaccines as it is the most clinically developed non-viral gene delivery system. Due to the precision of the LNPs to cautiously and productively deliver nucleic acids, it overcomes a major hindrance that was preventing the growth and the use of vaccines and genetic medicines. Researchers have found that the nanoparticles work much finer than other techniques such as T-tube mixing and thus encapsulating efficiencies of higher than 95% [14]. The following table (Table 2) discusses the advantages of mRNA vaccine's safety, efficacy, and production over the traditional vaccine.

4 mRNA Vaccine and Treatment

4.1 HIV

This is a disease where the virus attacks the white blood cells, and as a result, it weakens the immune system of the body. Therefore, it makes a person more vulnerable to other diseases and infections. It is spread during unprotected sexual intercourse or through sharing injection drug equipment. Current treatment approaches for HIV involve the use of antiretroviral drugs for antiretroviral therapy (ART) and pre-exposure prophylaxis (PrEP); it has played a vital role transforming AIDS from a life-threatening illness to a controllable life-threatening disease. But, its medications are high-priced, and the intake of its doses should be taken properly as it triggers side effects. Researchers have found that mRNAs with LNPs are capable of stimulating HIV antibodies. These antibodies recognize the targeted pathogen with the help of genetic material that instruct the immune cells and are able to encode complex antigens which is the key to HIV vaccine development. The biological study of broadly neutralizing antibody [bnAb] development, the development of an effective HIV vaccine is more complicated than the vaccines available in the present times. We have visualized a series of Env-based immunogens that will lead to B cell maturation in the direction of bnAb from germline precursor in germinal centers, via sequential immunization when carried and encoded as mRNAs (see Fig. 2). It is very crucial that a vaccine elicits several bnAb lineages which target distinct conserved sites on the Env glycoprotein since the HIV has the capability to escape from the antibody responses and may need multiple series of immunogens. In a recent experimental study, three HIV mRNA vaccines are introduced; they are BG505MD39.3 mRNA, BG505MD39.3 gpl151 mRNA, and BG505MD39.3 gpl151 CD4KO mRNA.

The major function of each of the vaccines is intended to demonstrate the spike protein which will be found on the surface of the virus (HIV) that allows the entry into human cells. Each experiment shows that the produced vaccines are highly related and stabilized proteins [17–20].

4.2 Rabies

This disease causes fatal inflammation of the spinal cord and brain which is caused by the zoonotic pathogens. There are two forms of rabies: furious rabies, which causes hyperactivity and hallucinations and paralytic rabies, which causes paralysis and coma. In the clinical trial, researchers had found the model antigen rabies virus glycoprotein (RABV-G) that could be used in the treatment of rabies virus. According to the results of previous trials on human beings using rabies glycoprotein as an antigen, it is established that lipid encapsulation protects the mRNA molecules in our body and improves the production of protective immune responses after two doses in a parallel manner to the deactivated control vaccine. This antigen had been clearly characterized and defined. Moreover, in phase 1 trials, it has been found that there is virtually 100% fatality outcome of rabies disease. This means that the volunteers taking part in the clinical trial had no history of rabies vaccination that will represent an immunologically unaffected population. During the vaccination procedure, the mRNA carries the information of RABV-G which is the rabies virus protein and will directly be injected to the muscle, and as a result, the vaccinated person's body will themselves produce the RABV-G protein which helps the immune system and recognizes the protein that triggers the immune response [21, 22].

4.3 Cancer

mRNA-based drugs have been revolutionized in such a way that it has been used as cancer therapy by discovering application in several kinds of anticancer activity such as immunomodulatory drugs, CAR cell therapies, and monoclonal antibodies. One of the important traits of mRNA is it exhibits several therapeutic benefits unlike other

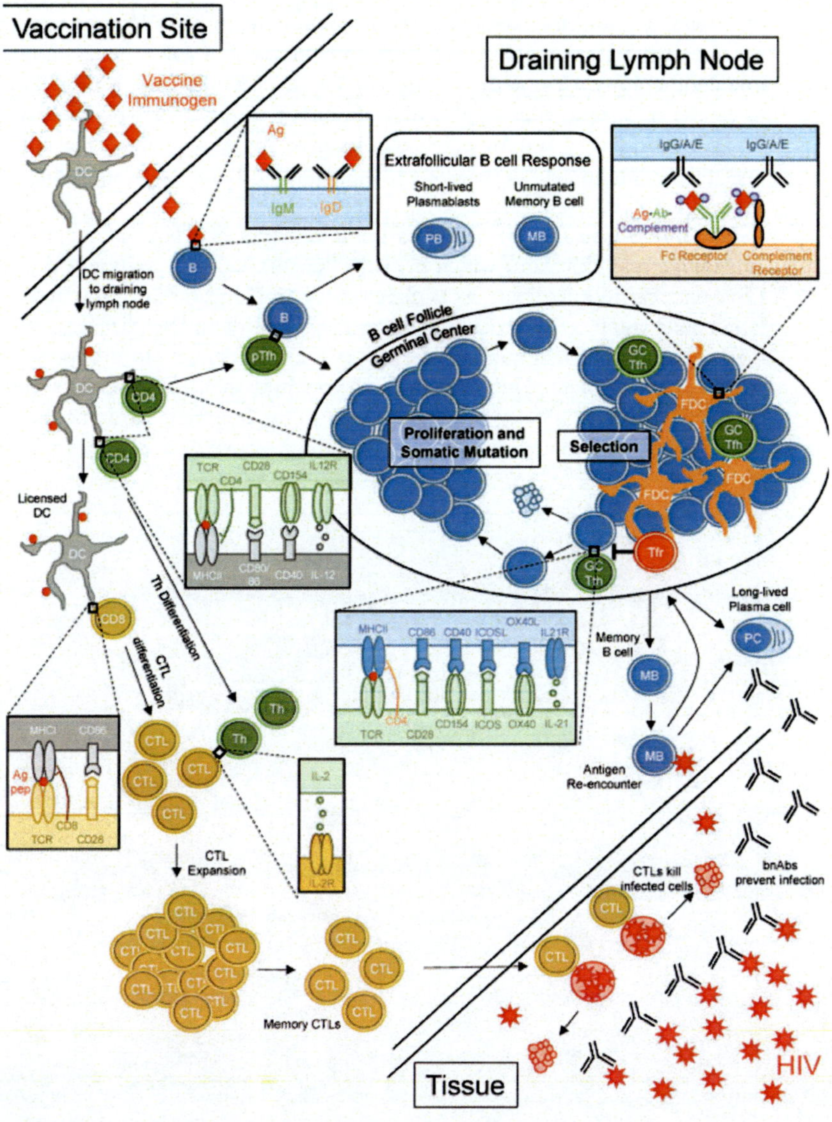

Fig. 2 Figure describes the humoral and cellular responses through vaccination [17] [Adapted from Zekun [18] (CC BY 4.0)]

functional biomolecules such as recombinant proteins and pDNA. As a result, mRNA vaccines allow us to explore the growth and treatment of a new generation of cancer immunotherapy drugs. Although mRNA-based vaccines have enormous potential in cancer treatment, their application became possible only with the existence of LNPs in its delivery system which enables diversity of new anticancer medications presently in preclinical development, and some of them are going through clinical trials.

The figure (see Fig. 3) describes the interaction of RNA aptamers (bispecific, antagonist, and chimeric) which helps in the survival and antitumor activity [23]. Lipid nanocarriers application has directed key problems for mRNA transfection into target cells and thereby protecting it from degradation in the extracellular compartment, in addition to facilitate cellular percipience and delivery to an appropriate intracellular compartment. Therefore, we can conclude that the advances in the design of mRNA using LNPs can be of great help in the treatment of numerous forms of cancer using immunotherapy [24].

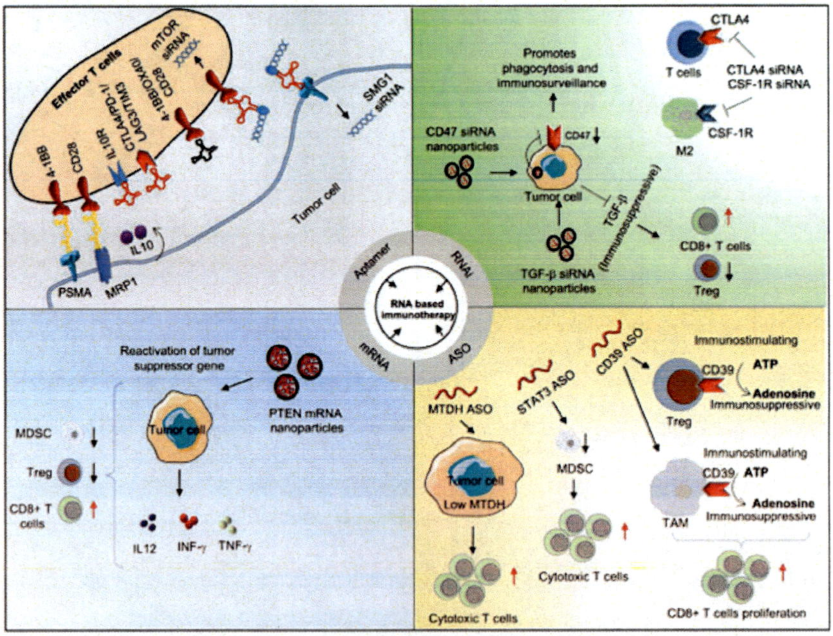

Fig. 3 RNA-based therapeutic diet for improving RNA-based immunotherapy [23] [Adapted from Poonam R [24] (CC BY 4.0)]

4.4 Coronavirus

Coronavirus (COVID-19) is a contagious disease caused by the SARS-CoV-2 virus. For this disease, the vaccines consist of composed mRNA strands encapsulated in LNPs. These mRNA vaccines were the first to receive 'emergency use authorization' and 'conditional approval' by EMA. The major benefit of mRNA vaccine is that it can be developed faster than the other type of vaccines. The mRNA can be developed within weeks after the identification of the protein antigen(s) and sequencing the subsequent gene(s). For the prevention of covid-19, mRNA vaccines are more effective than the viral vector vaccines as it doesn't generate immunity against the carrier. The LNPs also play a vital role in the development of the vaccines that enables successful freeze-drying, and as a result, the structures are exposed to stress. The lyophilization stabilizes the colloidal particles that should be incorporated in the formulation. The LNPs play an important role in the development of covid-19 vaccines; hence, it helps in transporting and protecting the mRNA effectively to target the cells. Thus, the use of LNPs is pushing boundaries for the drug delivery process. Due to the flexible layout, comparatively short-lived cytoplasmic presence, and normalized production procedures, mRNA vaccines play an important role amidst the pandemic [25, 26].The Fig. 4 shows the comparison of Moderna booster, Pfizer, and Johnson & Johnson's vaccine, whose efficiencies are 93%, 88%, and 71% respectively. We can observe the distinct efficiency gap between the m-RNA vaccines (Moderna booster and Pfizer vaccine) over viral vector vaccines (Johnson & Johnson's vaccine) [27].

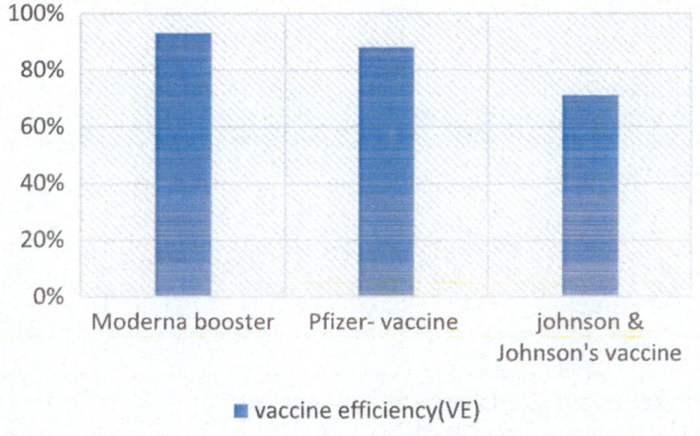

Fig. 4 Abovementioned graph shows the vaccine efficiency of the mRNA vaccine (Moderna booster and Pfizer vaccine) over viral vector vaccine (Johnson & Johnson's vaccine)

5 Conclusion and Future Scope

This review article outlines the overall aspect of the 'Development of mRNA vaccines with the aid of LNPs.' The following points are the summary of this review article as follows:

- It focuses on the manufacturing of the mRNA vaccine for different types of diseases such as HIV, rabies, cancer, and covid-19 with the help of LNPs which act as a protein carrier directly into the cells of the human body.
- One of the important traits of mRNA vaccines is that it has the capability to be used prophylactically to avoid the infection or disease and therapeutically to enhance immunity post-infection/disease. Since we know that the manufacturing of complex nanoparticle protein immunogens on a larger scale poses a significant challenge, strategically achievable and cost-effective production of mRNA raises the possibility of producing complex immunization.
- The introduction of LNPs in mRNA vaccines has been a revolutionary idea since it has the potential to encapsulate an assortment of cargoes like peptides, genetic payloads like siRNA, saRNA, and mRNA, and other small molecules.
- There are additional advantages of LNPs including their application in dermatology and in cosmetics. Researchers should emphasize to improve the optimization of the mRNA nucleotide composition for stability of LNPs along with the understanding of the environment to which mRNA is exposed in the core of the LNP that will help to rationalize the adjustment of the LNP structure for further integrity.

Thus, it will improve the scope of developing vaccines or drugs for any unknown diseases that we may have to face in the future.

Acknowledgements We express our sincere thanks to the Chancellor, Vice-presidents, the Management, Dean, HOD of the VIT-AP University, Andhra Pradesh for supporting a suitable International conference funding to attend, present the work in the ICANA conference and make this paper a successful one.

References

1. Wilson B, Geetha KM (2022) Lipid nanoparticles in the development of mRNA vaccines for COVID-19. J Drug Deliv Sci Technol 74:103553
2. Information on: https://covid19.who.int/
3. Information on: https://www.hhs.gov/immunization/basics/types/index.html#:~:text=There%20are%20several%20types%20of,Messenger%20RNA%20(mRNA)%20vaccines
4. Malabadi RB, Meti NT, Chalannavar RK (2021) Applications of nanotechnology in vaccine development for coronavirus (SARS-CoV-2) disease (Covid-19). Int J Res Sci Innov 8(2):191–198
5. Schlake T, Thess A, Fotin-Mleczek M, Kallen K-J (2012) Developing mRNA-vaccine technologies. RNA Biol 9(11):1319–1330

6. Tenchov R, Bird R, Curtze AE, Zhou Q (2021) Lipid nanoparticles—from liposomes to mRNA vaccine delivery, a landscape of research diversity and advancement. ACS Nano 15(11):16982–17015
7. Dolgin E (2021) The tangled history of mRNA vaccines. Nature 597:318–324
8. Jackson NAC, Kester KE, Casimiro D, Gurunathan S, DeRosa F (2020) The promise of mRNA vaccines: a biotech and industrial perspective. npj Vaccines 5(1):1–6
9. Information on: https://my.clevelandclinic.org/health/treatments/21898-mrna-vaccines
10. Reichmuth AM, Oberli MA, Jaklenec A, Langer R, Blankschtein D (2016) mRNA vaccine delivery using lipid nanoparticles. Ther Deliv 7(5):319–334
11. Hou X, Zaks T, Langer R, Dong Y (2021) Lipid nanoparticles for mRNA delivery. Nat Rev Mater 6(12):1078–1094
12. Ganesan P, Narayanasamy D (2017) Lipid nanoparticles: Different preparation techniques, characterization, hurdles, and strategies for the production of solid lipid nanoparticles and nanostructured lipid carriers for oral drug delivery. Sustain Chem Pharm 6:37–56
13. Information on: https://www.atascientific.com.au/the-role-of-lipid-nanoparticles-in-vaccine-development/
14. Anthony S, Priya D, Sivakumar K, Venugopal P, Sriram DK, George M (2021) Can mRNA vaccines turn the tables during the COVID-19 pandemic? Current status and challenges. Clin Drug Investig 41:499–509
15. Information on: https://www.vumc.org/viiii/infographics/how-does-mrna-vaccine-compare-traditional-vaccine
16. Ruiz-Fresneda MA, Ruiz-Pérez R, Ruiz-Fresneda C, Jiménez-Contreras E (2022) Differences in global scientific production between new mRNA and conventional vaccines against COVID-19. Environ Sci Pollut Res 29:1–13
17. Mu Z, Haynes BF, Cain DW (2021) HIV mRNA vaccines-progress and future paths. Vaccines 9(2):134
18. Information on: https://www.nih.gov/news-events/news-releases/nih-launches-clinical-trial-three-mrna-hiv-vaccines
19. Morris L (2021) mRNA vaccines offer hope for HIV. Nat Med 27(12):2082–2084
20. Information on: https://www.mlo-online.com/disease/hiv/article/21260738/hiv-vaccines-use-mrna-vaccine-technology
21. Aldrich C, Leroux–Roels I, Huang KB, Bica MA, Loeliger E, Schoenborn-Kellenberger O, Walz L, Leroux-Roels G, von Sonnenburg F, Oostvogels L (2021) Proof-of-concept of a low-dose unmodified mRNA-based rabies vaccine formulated with lipid nanoparticles in human volunteers, A phase 1 trial. Vaccine 39(8):1310–1318
22. Information on: https://www.who.int/health-topics/rabies#tab=tab_1
23. Pandey PR, Young KH, Kumar D, Jain N (2022) RNA-mediated immunotherapy regulating tumor immune microenvironment: next wave of cancer therapeutics. Mol Cancer 21(58):1–18
24. Guevara ML, Persano F, Persano S (2020) Advances in lipid nanoparticles for mRNA-based cancer immunotherapy. Front Chem 8:589959
25. Kulkarni JA, Thomson SB, Zaifman J, Leung J, Wagner PK, Hill A, Tam YYC, Cullis PR, Petkau TL, Leavitt BR (2020) Spontaneous, solvent-free entrapment of siRNA within lipid nanoparticles. Nanoscale 12(47):23959–23966
26. Schoenmaker L, Witzigmann D, Kulkarni JA, Verbeke R, Kersten G, Jiskoot W, Crommelin DJA (2021) mRNA-lipid nanoparticle COVID-19 vaccines: structure and stability. Int J Pharm 601:120586
27. Information on: https://www.yalemedicine.org/news/johnson-and-johnson-covid-booster

Metal–Organic Framework for Antibiotic Sensing Application

Krupa U. Patel, Dashrathbhai B. Kanzariya, and Tapan K. Pal

Abstract Due to the harmful environmental threat posed by antibiotics, there has been a significant increase in research interest in developing appropriate chemosensors for specific quantification and detection of antibiotics. Creating proper sensors is required because the excessive buildup of antibiotics inside the soil, caused by high human consumption and animal activity, might harm biological systems. Among the several sensing methodologies, the fluorescence-based technique has emerged as an alternative and efficient way to identify antibiotics. Numerous fluorescence sensors have been developed for the discriminative detection of antibiotics in solution and vapor phases. Recently, it has been observed that a novel category of hybrid materials, metal–organic frameworks (MOFs), is a promising material for the selective and sensitive detection of antibiotics. The MOF comprises metal ions (metal node or secondary building unit) with precisely articulated organic ligands. It possesses excellent qualitative structural features such as various topology, flexibility, highly stable porosity, and automatically located functional sites, which bestow them to perform as an efficient luminescent sensor. In this book chapter, we have methodically summarized the current developments using luminescent MOFs as fluorescence sensors for the discriminative detection of various antibiotics. Finally, this chapter describes this conclusive remark of this chapter.

Keywords Metal–organic frameworks (MOFs) · Antibiotics · Drug · Sensing

1 Introduction

The discovery of antibiotics at the beginning of the twentieth century caused a paradigm change in the study of microbiological sciences (diseases caused by microorganisms). This finding led to the development of a wide variety of antibiotics that are successfully used to treat on human/animal to stand against the communicable diseases.[1] Recently, the ingestion of antibiotics is dramatically increased and

K. U. Patel · D. B. Kanzariya · T. K. Pal (✉)
Department of Chemistry, Pandit Deendayal Energy University, Gandhinagar, Gujarat 382426, India
e-mail: Tapan.pal@sot.ac.in

N. M. Rao et al. (eds.), *Advanced Nanomaterials and Their Applications*,
Springer Proceedings in Materials 22, https://doi.org/10.1007/978-981-99-1616-0_12

is also widely utilized in producing healthcare items.[2] Antibiotics have accumulated excessively in waterbodies and ecosystems due to heavy consumption, excretion from human body, and incorrect disposal, ultimately affecting the ecosystem and people's health.[3] Each day, tons of antibiotics are generated, and every year tons of chemical contaminants are directly discharged into the atmosphere.[4] Additionally, the wastewater instantly removed from the industries which contains a substantial amount of antibiotics.[5] The regular consumption of antibiotics might lower the immunity and result in gastric, neurotoxic, allergy symptoms, genetic issues, and different tumors.[6,7]

Furthermore, the growth of microbial infections which developed may resist to nearly all antibiotic drugs, and this is very severe threat to public health.[8,9] To improve, the human health and safeguard for the environment, it is crucial to develop an efficient and trustworthy detection approach for residual analysis and detection of antibiotics, with excellent selectivity and sensitivity. Quite well-known, the sophisticated analytical methods can be used to identify antibiotics[10,11]; however, these methods sometimes suffer for routine analysis due to their inherent drawbacks such as time requirement and solution preparation, and due to their heavy weight, they cannot be used as a portable sensor.[12]

Luminescent metal–organic frameworks (LMOFs) are meticulously investigating for the recognition of ions, pollutants, and small molecules.[13,14] Recently, the MOFs have been emerged as a suitable sensor for the monitoring of antibiotics.[15] LMOFs possess various advantages like structural diversity, diverse functional groups, purposely implanted accessible active sites, porosity, guest interaction site and has the ability to show structure and property connection.[16] Yet, the monitoring of hazardous and biologically important organo-aromatics requires the capability to show ultralow-level detection, quick response time, and multicycle performances ability. The difficulty in creating a multisensory system with rationalizing the unique signal transduction is very imperative, which implies that the sensor should have the distinct fingerprints in the detection of desired analyte in presence of the other interfering analyte. More importantly, the visual recognition of analyte can be of important research motif for real-time application as it offers the naked eye recognition of analyte. Here, we have systematically portrayed the advancement in the detection of luminescent MOFs for the detection of various antibiotics.

2 Fluorescence Sensors for Sulfonamide Antibiotics Based on LMOFs

The first antibacterial agents that sparked in the revolution of antimicrobial medications were sulfonamides (SPA), also known as sulfa-drugs. The sulfamethazine (SMZ), sulfadiazine (SDZ), sulfameter (SM), sulfachlorpyridazine (SCP), sulfamethoxazole (SMX), sulfaquinoxaline (SQX), and sulfamonomethoxine

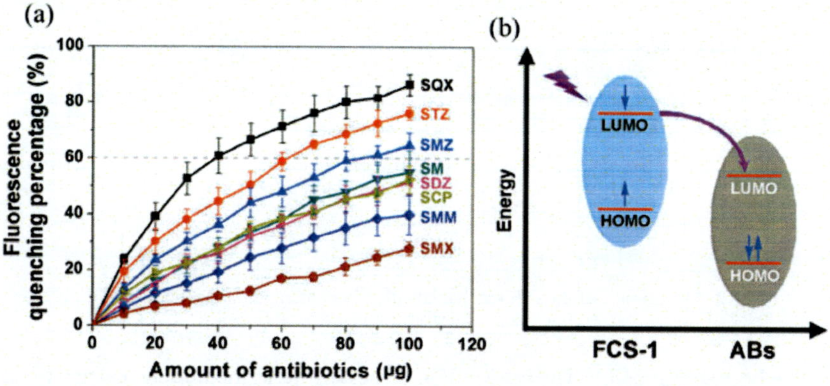

Fig. 1 a Luminescence quenching percentage of MOF in the existence of various drugs and **b** mechanism of releasing electrons from LUMO of MOF to LUMO of antibiotics

(SMM) are the different types of sulfonamide-based drug molecules. These sulfa-drugs hinder the bacterial manufacture of folic acid, a critical stage in cell development, acting as antibacterial agents and halting the bacterial growth. The SPA antibiotic would be an effective against bacterial activity. MOFs showed its efficacy in the recognition of SPA antibiotics. For example, Liu and co-worker reported the hydrothermal reaction among H_3L and $Zn(NO_3)_2{\cdot}6H_2O$ for the synthesis of highly crystalline luminescent MOF, $[Zn_3(3\text{-}OH)(HL)L(H_2O)_3]{\cdot}H_2O]_n$. This MOF was utilized to ultra-sensitive recognition of SPA antibiotics in wastewater (Fig. 1).[17]

In the solid state, the ligand H_3L and the MOF showed the maximum emission peak at 375 and 382 nm, respectively. The long-range electronic connection between the nearby organic linkers is stop which causes to show the higher fluorescence intensity for the MOF. With treatment of various SPA antibiotics, the luminescence intensity of MOF showed diverse quenching effect (Fig. 1a). For SQX and STZ, the luminescence intensity decline is highest, and these may be due to the strong binding energy with an order of 10^{-4} M^{-1} in an aqueous media. The presence of hydrophilic functional moiety on the MOF backbone involves the superior interaction with the antibiotics which allow the facile charge transfer and thereby reduces the fluorescence intensity of MOF. The PET effect operates here, transferring an excited electron from HOMO of probe (MOF here) toward the LUMO of antibiotics, which lowers MOFs emission intensity (Fig. 1b). The SCP antibiotics showed a limited fluorescence quenching efficiency because they did not interact through either π–π or hydrogen bonds. This MOF is a promising sensitive sensor for potential use in wastewater treatment since the strong sensing performance of MOF for sulfa-drugs was maintained over an extensive pH range of 3–9 and was unaffected even by the co-survival of several heavy metal cations.

Recently, it is noticed that the lanthanide-containing MOFs are excellent luminescence sensors used to detect sulfonamide antibiotics. As a promising probe, for accurate and irreversible identification of the antibiotic SMZ, a 2D Eu-MOF has

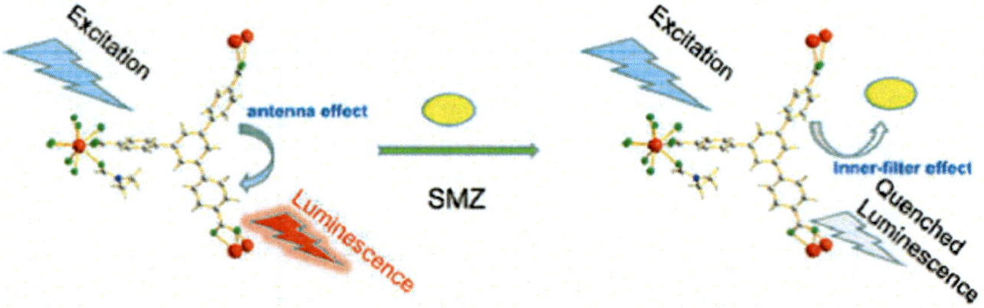

Fig. 2 Occurrence of the IFE mechanism clearly explained the quenching of Eu-MOF

been developed.[18] The Eu-MOF in dispersed water medium revealed a typical well-resolved europium red emitted as a result of the efficient energy exchange from the sensitizer H_3BTB to the Eu(III) ion (Fig. 2). The stepwise addition of SMZ to the water dispersed medium of Eu-MOF, the luminescence intensity of the MOF is gradually reduced. The operation of IFE mechanism results in the reduction of luminescence intensity of the MOF.

3 Fluorescence Recognition on LMOFs with Antibiotics that Include Nitro

It has been observed that the antibiotic consisting imidazole and nitro moiety is very effective to control the fungal infections. Some of the antibiotics of this class are dimetridazole (DTZ), metronidazole (MDZ), and nitroimidazole (NDZ). These medicines would damage the DNA of microbial cells, which prevents the creation of nucleic acids. The other nitro-containing antibiotics such as furantoin, furazone, and furazolidone (FZD) are listed under the nitrofuran derivatives. Even they have the good therapeutic value, but the excessive usages of these antibiotics can causes several adverse effect. Here, we also showed the utilization of several LMOFs for their sensing capabilities of various nitro-containing antibiotics.

Li and co-worker reported the hydrothermal reaction between $Ln(NO_3)_3{\cdot}6H_2O$ with 2,3′-H_2oba to create an Eu-MOF, which demonstrated the specific fluorescence-sensing capability for the recognition of antibiotic MDZ in an aqueous medium (Fig. 3).[19] When MDZ was added incrementally, the characteristic red emission of Eu-MOF was gradually reduced. However, the antibiotics do not contain any nitro group did not affect much on the luminescence intensity of the Eu-MOF. The electron-rich complex Eu-MOF transfers its energy to the electron-deficient MDZ is believed to cause fluorescence quenching.

Jing et al. recently derived an excellent luminescent zinc-MOF, Zn-MOF for the detection of nitro-containing antibiotics.[20] The hydrothermal interaction of Zn $(NO_3)_2{\cdot}6H_2O$ with 4,4′-bpy and oba in DMA to delivered the Zn-MOF. In comparison the luminescence intensity of ligands, 4,4′-oxybis (benzoic acid), and 4,4′-bipyridine,

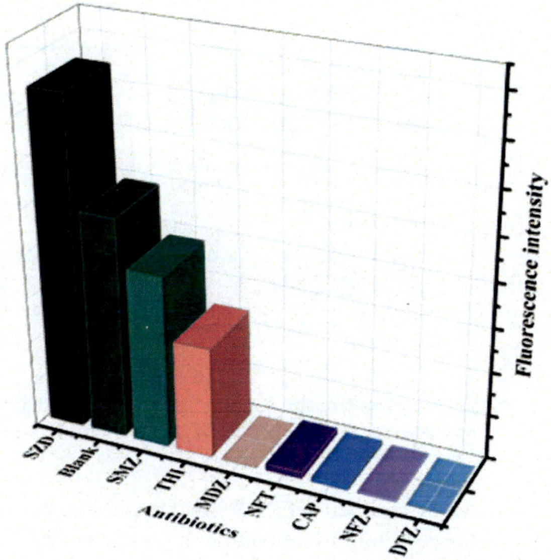

Fig. 3 This bar diagram illustrates the selective fluorescence quenching of Eu-MOF by DTZ

the Zn-MOF showed an intense emission band at 445 nm, and this is owing to the LMCT.

The luminescence screening of the MOF with diverse antibiotics, it was revealed that the emission intensity is selectively reduces in the attendance of metronidazole (MET) (Fig. 4a). Other antibiotics and interfering species like amino acids, anions, cations, and small organic compounds have barely shown the fluorescence quenching effects of sensor under the same conditions. Further, the Zn(II)-MOF is a practical fluorescence probe for the recognition of organic pollutant in order to measurement of water quality. The MOF materials become reusable at least five cycles for the recognition of MET with nominal change in the recognition ability (Fig. 4b).

Interestingly, a luminescence mixed membrane MOF composite, Eu-TDC MMMs was synthesized from $[Eu_2(TDC)_3(CH_3OH)_2(CH_3OH)]$ with a glue (502 tubes of glue from the Evo-bond product (EVOB).[21] The excellent stability and remarkable selectivity of Eu-TDC MMMs endow the discriminating recognition of nitroimidazole antibiotics (NIABs) in an aqueous media. The observed quenching behavior of Eu-TDC MMMs is the occurrence IFE mechanism.

A two-fold luminous Zn-MOF, $[Zn_4O(BCTPE)_3]$, was produced by using AIE inapt tetra-phenyl ethylene (TPE)-based dicarboxylic acid with the zinc ion. Due to the AIE character of the embedded TPE ligand, the MOF demonstrated typical microporous behavior with highly emissive signature under heating conditions.

When antibiotics such as NFZ and MDZ were gradually added to $[Zn_4O(BCTPE)_3]$, the initial emission intensity of MOF was significantly lowered, with a quenching frequency 93 and 50%, respectively. The luminescence quenching

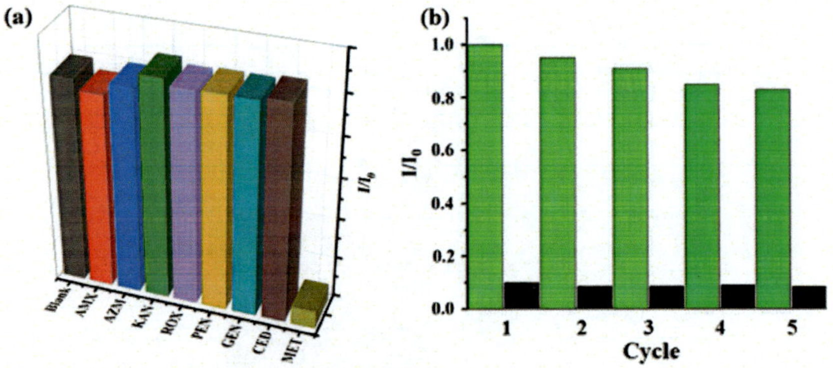

Fig. 4 **a** Fluorescence intensity ratio of I/I_0 of MOF upon addition of various antibiotics; and **b** reversible sensing ability of MOF to antibiotic MET over five cycles

would ascribed to FRET effect because of the extensive overlap between the absorption spectra of both antibiotics with the emission profile of MOF. This results the intermolecular charge transfer from the probe to the analytes, afforded the significant quenching of the sensor. This MOF showed the recognition ability up to three cycles (Fig. 5). In another study, a p-electron-rich terphenyl-tetracarboxylic acid (H_4ptptc) was converted into an alkali-resistant Zn-MOF, $(BMI)_2[Zn_3(ptptc)_2]$ under combined hydro/solvothermal and ionothermal reaction conditions. The authors have created a novel crystallization environment to increase the crystal yield by dissolving combining organic and inorganic substrates in [BMI]Br ionic liquid.[22] By π–π stacking interactions, the ligand H_4ptptc, which has delocalized p-electron, enriched the luminous characteristics and increased the stability of the Zn-MOF framework. The metal–oxygen coordination connection in MOF is very strong and tightly packed the frameworks which provide a rigid matrix that lessens the non-radiative absorption. Its trinuclear Zn(II) clusters in MOF are coordinated only with carboxylate groups of the ptptc ligand, and highly disordered $[BMI]^+$ cations fill the crystal gaps. The structural network is shown its chemical stability in water and in alkaline media. The addition of antibiotics, the emission intensity of MOF is significantly reduced (Fig. 6).

In particular, the antibiotics NZF and DTZ have higher fluorescence quenching efficiencies of 97 with 90%, respectively. The mechanism of sensing is established with the assistance of DFT analysis. A highly luminescent Tb-MOF [Tb(TATAB)(H_2O)] is reported by the Li group.[23] In the aqueous medium, this MOF demonstrated the recognition of nitroimidazole (ODZ, MDZ, DTZ, and RDZ) through fluorescence "turn-off" effect.

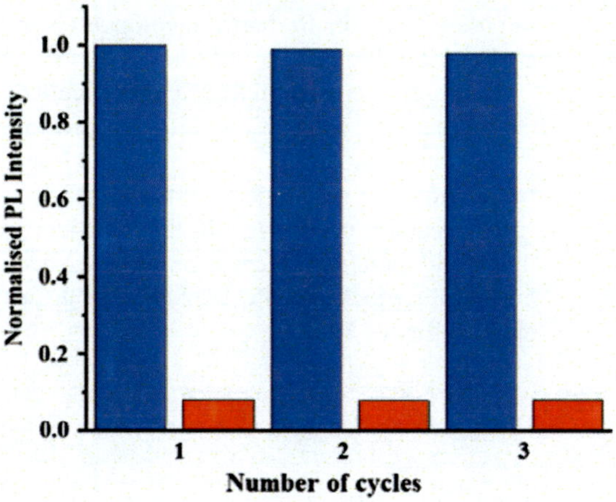

Fig. 5 Recyclability test of the MOF for the detection of antibiotics (the red and blue bars signify the fluorescence intensity of MOF after adding 65 ppm nitrofurazone and sole luminescence intensity, respectively)

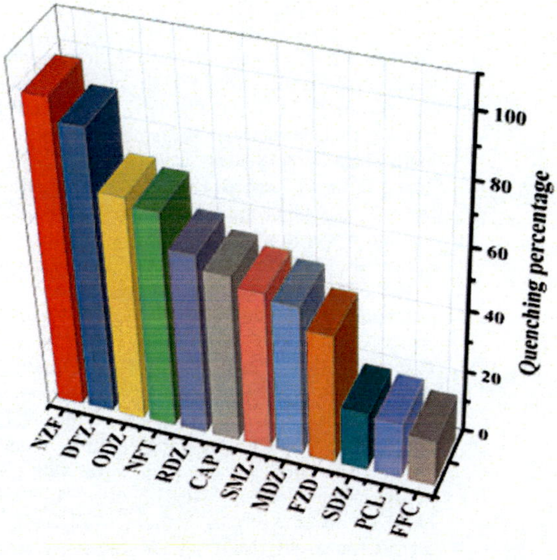

Fig. 6 Luminescence quenching percentage of MOF in the existence of various antibiotics

4 Conclusion

The LMOFs can efficiently detect structurally and electrically various antibiotics. We thoroughly demonstrated the different LMOFs have shown their detection capability through either luminescence enhancement or quenching effect. The diverse important features such as guest functional site, porous nature, high stability, and structural flexibility of the MOFs would endow the MOF to behave as a luminescent sensor. Over the past few years, many studies have been reported on LMOFs as possible sensors for harmful compounds, biologically significant organisms, and environmental toxins.

However, even though MOFs have excelled in the detection of diverse antibiotics, but for real-time application, further improvement is necessary. For a reliable sensor, it is imperative that the materials should be constructed from low-cost materials. In addition, the sensitivity, ultra-selectivity, and reusability of the material is also crucial features needs to be further consideration. The judicial selection of organic linker and metal ion, there is an ample space available for the development of the new MOFs which can effectively handle the technical issues to address the real-time applicability of MOFs.

Acknowledgements KUP and DBK acknowledge SODH and PDEU, respectively, for scholarship. TKP gratefully acknowledges financial support received from the SERB (TAR/2021/000090).

Notes

1. Calderon CB, Sabundayo BP (2017) Antimicrobial classifications. Antimicrobial classifications: drugs for bugs. CRC Press, Taylor & Frances Group. ISBN 978-0-8247-4100-6.
2. Fan M, Sun B, Li X, Pan Q, Sun J, Ma P, Su Z (2021) Highly fluorescent cadmium based metal–organic frameworks for rapid detection of antibiotic residues, Fe^{3+} and $Cr_2O_7^{2-}$ ions. Inorg Chem 60:9148–9156.
3. Zhang C, Lai C, Zeng G, Huang D, Yang C, Wang Y, Zhou Y, Cheng M (2016) Efficacy of carbonaceous nanocomposites for sorbing ionizable antibiotic sulfamethazine from aqueous solution. Water Res 95:103–112.
4. Byrne MK, Miellet S, McGlinn A, Fish J, Meedya S, Reynolds N, van Oijen AM (2019) The drivers of antibiotic use and misuse: the development and investigation of a theory driven community measure. BMC Public Health 19:1425.
5. Hanna N, Sun P, Sun Q, Li X, Yang X, Ji X, Zou H, Ottoson J, Nilsson LE, Berglund B, Dyar OJ, Tamhankar AJ, Lundborg CS (2018) Presence of antibiotic residues in various environmental compartments of Shandong province in eastern China: its potential for resistance development and ecological and human risk. Environ Int 114:131–142.
6. Zhao D, Liu X-H, Zhao Y, Wang P, Liu Y, Azam M, Al-Resayes SI, Lu Y, Sun W-Y (2017) Luminescent Cd(II)–organic frameworks with chelating NH_2 sites for selective detection of Fe(III) and antibiotics. J Mater Chem A 5:15797–15807.
7. Zhu XD, Zhang K, Wang Y, Long WW, Sa RJ, Liu TF, Lü J (2018) Fluorescent metal–organic framework (MOF) as a highly sensitive and quickly responsive chemical sensor for the detection of antibiotics in simulated wastewater. Inorg Chem 57:1060–1065.
8. Nikaido H (2009) Multidrug resistance in bacteria. Annu Rev Biochem 78:119–146.
9. Tenover FC (2006) Mechanisms of antimicrobial resistance in bacteria. Am J Med 119:S3–S10.
10. Luiz MMA, Vidal JLM, Gonzalez RR, Frenich AG (2008) Multi-residue determination of veterinary drugs in milk by ultra-high-pressure liquid chromatography–tandem mass spectrometry. J Chromatogr A 1205:10–16.
11. Cinquina AL, Longo F, Anastasi G, Giannetti L, Cozzani R (2003) Validation of a high-performance liquid chromatography method for the determination of oxytetracycline, tetracycline, chlortetracycline and doxycycline in bovine milk and muscle. J Chromatogr A 987:227–233.

12. Moreno-Gonzalez D, Lara FJ, Jurgovská N, Gá L, García-Campañ AM (2015) Recent advances in luminescent metal–organic frameworks (LMOFs) based fluorescent sensors for antibiotics. Anal Chim Acta 891:321–328.
13. Patel N, Shukla P, Lama P, Das S, Pal TK (2022) Engineering of metal organic frameworks (MOFs) as ratiometric sensors. Cryst Growth Des 22:3518–3564.
14. Goswami R, Pal TK, Neogi S (2021) Stimuli-triggered fluoro-switching in metal–organic frameworks: applications and outlook. Dalton Trans 50:4067–4090.
15. Goswami R, Mandal SC, Seal N, Pathak B, Neogi S (2019) A Antibiotic-triggered reversible luminescence switching in amine-grafted mixed-linker MOF: exceptional turn-on and ultrafast nanomolar detection of sulfadiazine and adenosine monophosphate with molecular keypad lock functionality. J Mater Chem 7:19471–19484.
16. Kitaura R, Noro S-I, Kitagawa S (2004) Functional porous coordination polymers. Angew Chem Int Ed 43:2334–2375.
17. Zhu XD, Zhang K, Wang Y, Long WW, Sa RJ, Liu TF, Lu J (2018) Fluorescent metal–organic framework (MOF) as a highly sensitive and quickly responsive chemical sensor for the detection of antibiotics in simulated wastewater. Inorg Chem 57:1060–1065.
18. Ren K, Wu S-H, Guo X-F, Wang H (2019) Lanthanide organic framework as a reversible luminescent sensor for sulfamethazine antibiotics. Inorg Chem 58:4223–4229.
19. Li JM, Li R, Li X (2018) Construction of metal–organic frameworks (MOFs) and highly luminescent Eu(III)-MOF for the detection of inorganic ions and antibiotics in aqueous medium. CrystEngComm 20:4962–4972.
20. Wang J, Zha Q, Qin G, Ni Y (2020) A novel Zn(II)-based metal–organic framework as a high selective and sensitive sensor for fluorescent detections of aromatic nitrophenols and antibiotic metronidazole. Talanta 211:120742–120750.
21. Li C, Zhang F, Li X, Zhang G, Yang Y (2019) A luminescent Ln-MOF thin film for highly selective detection of nitroimidazoles in aqueous solutions based on inner filter effect. J Lumin 205:23–29.
22. Qin JH, Huang YD, Shi MY, Wang HR, Han ML, Yang XG, Li FF, Ma LF (2020) Aqueous-phase detection of antibiotics and nitroaromatic explosives by an alkali-resistant Zn-MOF directed by an ionic liquid. RSC Adv 10:1439–1446.
23. Wei JH, Yi JW, Han ML, Li B, Liu S, Wu YP, Ma LF, Li DS (2019) A Water-stable terbium(III)–organic framework as a chemosensor for inorganic ions, nitro-containing compounds and antibiotics in aqueous solutions. Chem Asian J 14:3694–3701.

Metal Organic Framework (MOF)-Based Membranes for Separation Applications

Dashrathbhai B. Kanzariya, Krupa U. Patel, Rudra Desai, and Tapan K. Pal

Abstract Hybrid organic–inorganic porous crystalline materials, i.e., metal organic frameworks (MOFs), are constructed of bridging organic ligands and metal connectors. However, as synthesised MOFs, they are in powder or crystalline form, which is incompatible with real-field applications. To make MOFs most efficient for real field application, it is necessary to embed or load MOF materials with other support materials to use them as in desired applications. MOF-based mixed-matrix polymeric membranes (MMPMs) are taking keen interaction in the field of membrane-based separations because of their high stability and multicyclic ability. MOFs are widely used in gas separations, waste water treatment, marine oil–water separations, desalination, and many more applications. In this chapter, we addressed MOFs-based MMPM-driven separation processes in various areas. We also described the recent report findings that give vital insights into these membrane materials, particularly at separation and their mechanisms to give better insight of MOFs interactions with other foreign analytes, which leads to finding more sustainable and efficient solutions to the current separation technologies.

Keywords Metal organic frameworks (MOFs) · Membranes · Separations · Gas separation · Waste water treatment

1 Introduction

For any living species on earth, water is an essential resource and the foundation of their life.[1] Despite water covering 72% of earth crust, just 0.5% is freshwater. Due to exponential population expansion, climate change, and water pollution, water supplies are dwindling.[2] It is of the utmost importance to address the issue of water shortage immediately. Particularly, the reduction of water pollution is expected to be one of the primary ways for addressing water scarcity.[3] In the water treatment protocol, separation technology became most prominent and effective solution tool.

D. B. Kanzariya · K. U. Patel · R. Desai · T. K. Pal (✉)
Department of Chemistry, Pandit Deendayal Energy University, Gandhinagar, Gujarat 382426, India
e-mail: tapan.pal@sot.pdpu.ac.in

N. M. Rao et al. (eds.), *Advanced Nanomaterials and Their Applications*, Springer Proceedings in Materials 22, https://doi.org/10.1007/978-981-99-1616-0_13

Membrane separation has attracted considerable interest because it is energy efficient and having ability of the membrane pores to pick the appropriate molecules that allow them to permeate without phase change or chemical addition.[4] Therefore, membrane separation, which has been viewed as a promising strategy for solving energy sector issues by considering the real environmental challenges, has become the most considerable issues during the past few decades. Membranes provide the numerous applications in the areas like separation, purification, and concentration of various components of a desired liquid input by means of discriminatory separation, although permeability and selectivity are incompatible.[5] To obtain promising outputs in terms of permeation and more selective results, scientist can not only tune the pore shape and size by introducing the new loading materials in it but also it is indeed requirement to replace a polymeric substrate for the betterment of the desired process. The composite membrane eases the trade-off between permeation and discrimination of particles by integrating the benefits of the two elements (e.g. the favourable polymer phase flux and the inorganic phase selectivity). Due to the poor agreement between the loading material and the polymeric materials (e.g. phases repulsion may diminish the porosity of the polymer), the creation and design of innovative loading material are essential.[6] Membrane materials play key role in achieving selectivity and high permeation towards the impurity refinement.[7] Therefore, it is necessary to identify an appropriate substance that can perform dual role of fillers and material for producing excellent membranes.

Metal organic frameworks (MOFs) are constructed of bridging organic ligands and metal connectors. MOFs are highly ordered structure having high surface area, unusual porosity, and tuning of pores and surfaces by altering the functional group that can facilitate the wider range analytes access.[8] In past few years, varieties of MOFs have been synthesised and widely employed in the various fields like storage of gaseous molecules, adsorption, separation, fluorescence applications, and many other water treatment applications.[9] MOF-based material is having high structural properties and showed great separation performance which instigated researchers to fabricate the composites to tackle numerous separation field challenges. MOF-based membrane can be synthesised by in situ growth technique, the seeding method, the layer-on-layer growth approach, and the electrochemical synthesis method. MOF-based mixed-matrix polymeric membranes might be synthesised with the help of fine dispersion of MOF particles throughout the polymer solutions.[10]

2 MOF: Synthetic Strategies

MOF-based polymeric membrane composites are found to deliver their optimum performance in separation techniques. And for that, it is indeed requirement for the MMPM materials consisting the water-stable MOF for the long-term stability of MMPM. MOF membrane design requires the optimal mixture of MOF in the polymeric materials during the membrane fabrication. This chapter emphasises the design methodologies for MOF-based membranes, mainly substrate support-based membranes and MMPM-loaded MOF composite membranes.

2.1 Substrate-Supported MOF Membranes

The MOF support-based membrane synthetic procedure mainly includes the straight growth of MOF particles on desired porous support, and the other one consists of secondary growth of MOF particles on substrate sites. For more detailed description straight growth as seen in Fig. 1a, nucleation and growth take place on support layer simultaneously. In a nutshell, straight growth can be observed on modified or unmodified supports. In this regard, Kang et al. created a chiral MOF membrane on a nickel net using a direct growth technique. Particularly, the substrate material in this case is nickel mesh was kept in an autoclave and then allowed to react with an appropriate concentration of an organic ligand. MOF crystals initially grew around the nickel mesh, and then continued to grow on each other over time. A variety of diol isomers was used to demonstrate the high separation efficiency. The support layer is having weaker bonding with substrate that requires chemical modification of support material to improve heterogeneous nucleation and directs the MOF membrane-based growth. Zhu et al. altered surface layer by biocomposite polydopamine (PDA) to synthesise high-crystalline ZIF membrane for the desalination applications.[11] For that, Zhu and team grafted PDA layer on support layer with the help of α-Al_2O_3 porous disc and dopamine at 20 °C for 20 h, resulting in a membrane with no fractures, pinholes, or other flaws. This indicates the favourable interaction and anchoring ZIF with surface of support, which facilitated nucleation and in situ growth of membranes. Additionally, a counter-diffusion in situ single-step approach which may be employed to create such membranes more easily and conveniently has been reported.[12] For secondary growth, the membranes are connected to the seed crystals of the MOFs; it is also an in situ approach (Fig. 1b). Secondary growth is not like primary growth, but it favours to regulate ultimate orientation to obtain high-dense membrane with no fractures or intergranular gaps.

2.2 MOF Composite Membranes

MOF-based composites fabricated by incorporating MOF particles into the polymeric materials to form MOF composite membranes. MOF loading into the polymeric matrix makes membrane more robust and introduces thermal strength into it. It also facilitates the permeate transport and provides better selectivity for the particular membranes.[13]

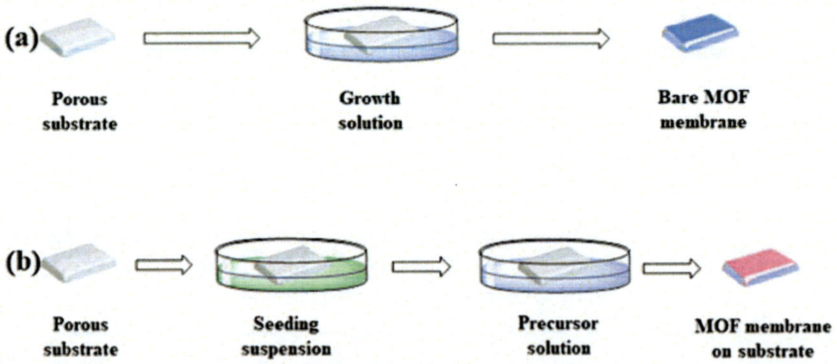

Fig. 1 **a** A schematic representation of straight growth of MOF and **b** schematic illustration of indirect in situ growth of MOF-based membranes of substrate layer

2.3 MOF Synthesis Methods

MOF structure synthesis can be done by the very renowned classical methods like,

1. Solvothermal/hydrothermal methods.
2. Ultrasound and microwave methods.
3. Mechanochemistry method.
4. Electrochemical methods and many other methods can be used to form MOF structures.

The developed MOFs can be used as filler materials in polymeric membrane formation process. Zhu et al. successfully prepared ZIF-8 nanoparticles (NPs) at room temperature (RT), and then the interfacial polymerisation techniques were utilised to fabricate the ZIF-8/PAN thin-film nanocomposite (TFN) to get optimum nanofiltration results.[14] Abbasi and co-workers developed HKUST-1 via ultrasound method.[15] Serra et al. produced NH_2-MIL-101(Cr) and chitosan polymeric matrix to get enhanced nanofiltration by solvothermal method[16] (Fig. 2).

2.4 MOF-Based MMPM Synthesis

The MMPM composite casting can be done by simply combining the two or more inorganic or organic–inorganic materials together to form new molecular hybrids. MOF-based MMPM is simple and cost-effective solution to scale up the synthetic method. MOF-based MMPMs are long-term stable and easy-to-fabricate kind of materials that can be achieved by two primary blending methods:

1. Blending on substrate
2. Blending without substrate.

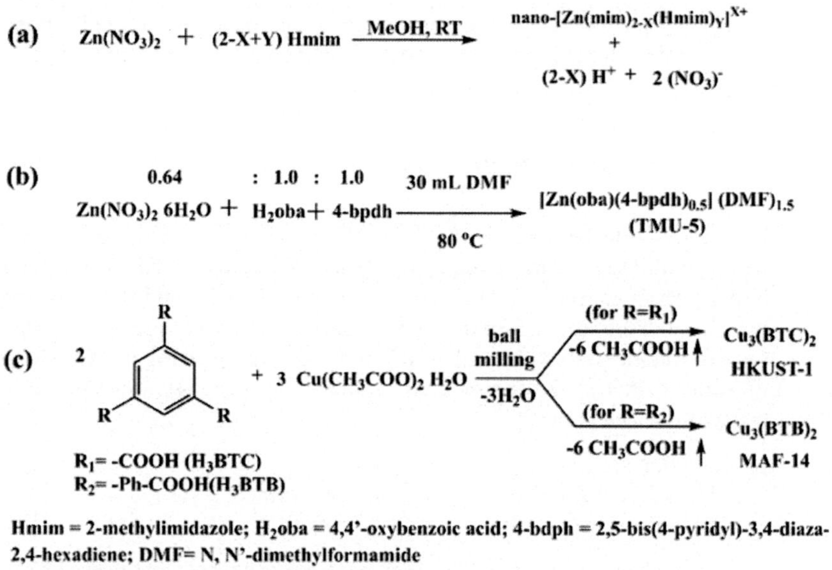

Fig. 2 **a** Solvothermal synthesis of ZIF-8 nanocrystals; **b** ultrasound-assisted TMU-5 synthesis, and **c** mechanochemical approach for HKUST-1 and MAF-14 synthesis

The blending on substrate method consists of following steps:

1. Mixing (MOF + solvent + polymer)
2. Solution casting on porous support
3. Removal of solvent and desiccation.

The blending without substrate method is nearly simpler to substrate-based method. For this method, nanoporous support is removed from the substrate. And the membrane thickness is set as per requirement (usually between 18 and 30 mm) to get high mechanical strength and great permeability membrane. Flyagina et al. reported PVDF@ZIF-90 produced MMPM without support layer,[17] and Benzaqui and coworkers reported PEG-g-ZIF-8/PVA MMPM composite made up of PEGylated ZIF-8 NPs and polyvinyl alcohol (PVA) for significant enhancement in selectivity and high flux.[18]

3 MOF-Based Membranes in Separation Applications

Based on the exceptional features of MOFs, researchers began to focus on the manufacture and use of MOF-based MMPM in recent years. In the last decade, a number of outstanding MOF-based MMPMs have been fabricated and widely utilised in variety of separations.[19] This section mainly focused on MMPM applications for gas mixture refinement, pervaporation, desalination, and oil–water treatment.[20]

3.1 Membrane Pervaporation

Pervaporation techniques can be defined as a method involving the separations of liquids with the help of membrane permeation and vaporisation of that particular liquid. Pervaporation facilitates the greater approach to the liquids that does not easily pass through the simple filtration. Pervaporation consists of mainly three processes, i.e. adsorption, dissolution, and diffusion rates of the given mixtures by dissolution–diffusion model in desired membrane layer.[21] Pervaporation can be divided into further three categories:

1. Dehydration
2. Volatile organic compounds (VOCs) refinement
3. Organic–organic solvent mixture separation.

In recent years, a great deal of study has been conducted on the use of ZIF hybrid membranes in pervaporation.

3.2 Desalination

Water shortages are becoming global challenges due to the rising pure water demand increases the water pollution, and 97% of the earth's crust having water is mainly from marine water resources, which must be immediately created and put to use. MOFs can have pore sizes as tiny as 0.2 nm or as big as 10 nm, allowing the usage of MOF MMPMs in both desalination applications and nanofiltration procedures as well as membranes for reverse osmosis (RO). Desalination covers the refinement of pure water molecules from the bulk salty sea water having numerous salt molecules dissolved in it. The research group of Huang created a dense and renewable Using bioexcitation, a ZIF salt membrane was created for the first time and polydopamine modification on the ceramic surface to facilitate desalination.[22]

3.3 Gas Separation

MOF membranes were initially utilised for gas separation. In addition, the key to achieving successful gas separation is selecting MOF membranes with the proper pore size with respect to the varying sizes of the gas molecules. Currently, alkene/alkane separation, hydrogen separation and refinement, and CO_2 separation are the primary focuses of gas separation.[23]

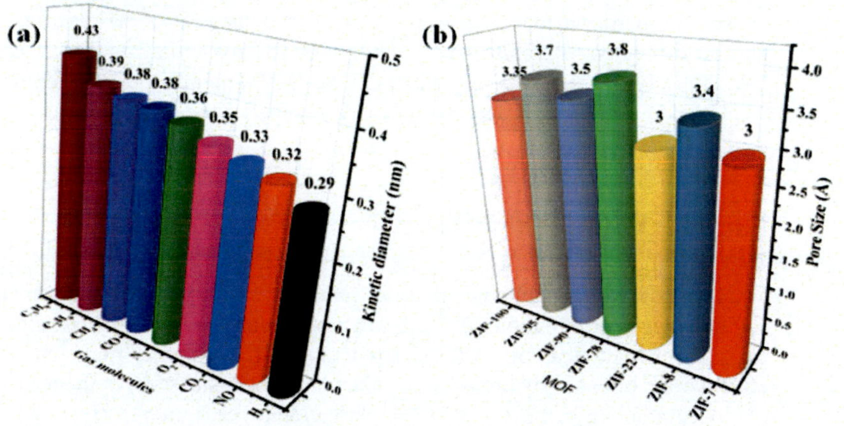

Fig. 3 **a** Kinetic diameter of various gas molecules and **b** ZIF-based MOF pore size

- Hydrogen separation and refinement

Hydrogen refinement is the most discussed topic in the field of MOF-based gas separation. Commercial manufacture of hydrogen energy, CO_2, N_2, and CH_4 are frequently combined with other tiny molecular gases, such as CO_2. Therefore, hydrogen refinement needs proper attention to completely remove hydrogen from the bulk pool of other gases. And this might be accomplished using MOF-based MMPMs with the appropriate pore size within it. Hydrogen's kinetic diameter is less than that of typical molecular gases, as shown in Fig. 3a. The membrane materials in the ZIFs class have the properties of molecular sieves similar to zeolites, as well as thermal and chemical stability. Due to their uniform microporous nature, they have lately found widespread usage in H_2 separation and purification. The ZIF membrane is used in H_2 separation that can be seen in Fig. 3b.[24]

- Separation and purification of CO_2

Carbon dioxide refinement with the help of MOF MMPMs is primarily covering the separation of CO_2/H_2, CO_2/N_2, and CO_2/CH_4 which are crucial linkages in the processes of fuel gas, natural gas, and hydrogen gas refinement. M-MOF-74 with 1.1 nm pore size contains free metal site and a very high carbon dioxide adsorption capability.[25] Kim et al. utilised the seed growth approach for fabrication of a continuous and full Ni-MOF-74 membrane having much greater carbon dioxide adsorption capability and membrane transmission followed by the mechanism of surface diffusion. Membrane separation ratios for N_2/CO_2, H_2/CO_2, and CH_4/CO_2 were larger than their respective Knudsen diffusion constants[26]: 9.1, 3.2, and 3.0, respectively.

- Alkenes/alkanes separation

As essential raw materials in the fossil fuel-based petroleum sector, alkenes (ethylene and propylene) are in high demand. The small alkanes (C_nH_{2n+2}) are having similar physical properties to that of alkenes (C_nH_{2n}) so that they can coexist in a gas mixture without difficulty. Therefore, the separation of alkenes and alkanes is essential for obtaining additional olefins. Currently, ZIF-8 MMPM membranes are predominantly employed in propylene/propane (C_3H_6/C_3H_8) separation.[27]

3.4 Oil–Water Separation

The cost-effective and reliable oil–water separation solutions are in demand to overcome rising water pollution. MMPMs are made up of polymeric matrix with desired MOF component loading in the desired ratios. The polymer materials are cost-effective and offer steady film-forming characteristics. And some of them are capable of biodegradation. In most of MMPM fabrication process, frequently used polymers are cellulose acetate (CA), polyvinylidene difluoride (PVDF), polysulphone (PSF), polyacrylonitrile (PAN), and polyethylene terephthalate (PET).[28] MMPMs of MOFs are utilised in oil–water refinement technology based on their structures and accessible functional groups. Oily waste water treatment process faces severe challenges in terms of membrane blockage because of their hydrophobic or oleophilic surfaces. Hydrophobic membrane surface may clog due to the deposition of oil particles on the membrane surface, and ultimately, flux loss is observed and it can be fixed by introducing filler materials.[29]

MOF-derived MMPM-based separation can be achieved by certain driving force and the binding affinity of the other analytes with membrane layer. Strong affinity and weaker affinity with the membrane layer construct the concentration gradient in liquid phase separation which can facilitate the molecule transport in the oily water, and it is referring to solid surface wettability. MMPMs also discriminate the molecules based on their particle size with respect to the accessible pores within the membrane layer. The larger molecules do not have access to the smaller membrane pores, and ultimately, they refined in the process. In addition to that, the oleophilic membranes allow hydrophilic particles through it and vice versa (Fig. 4). [24]

Qiu et. al. designed UiO-66-coated steel mesh[30] to obtain hydrophilic and underwater super-oleophobic nature with more than 99% efficiency and impressive permeation capability of 12.7×10^4 L m^{-2} h^{-1}. Their group also got optimum separation performance in case of ZIF-based membranes.[31] Yang and coworker[32] attempted in situ growth of natural fibre-based ZIF-8 nanocrystals for polydimethylsiloxane (PDMS) coating. The natural fibre-coated cotton/ZIF-8@PDMS fabric having self-cleaning properties was then employed oil–water separation applications and achieved separation rate of 95%.

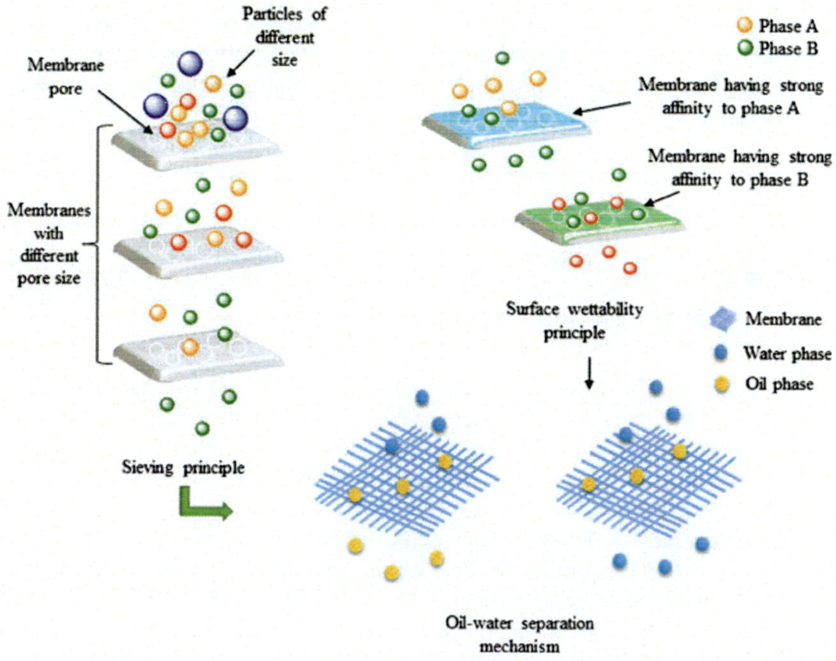

Fig. 4 Schematic representation of oil–water separation

4 Membrane-Based Separation Versus Adsorption Techniques

The MOF-based adsorbent materials are widely used in the varied range of purification techniques. The adsorbent materials have different kinds of affinity towards the target analytes and so that it can bind or trap the moiety to observe desired results. And according to their degree of affinity towards the contaminants, the removal results are observed. In the literature, a variety of adsorption materials are available like carbon-based materials, various nanoparticles, biomass, and many other materials which are widely employed in the adsorption process. In recent years, MOF materials are also explored for the various contaminant removals. Li et al. synthesised UiO-66-based zirconium MOF by doing some post-synthetic work to introduce the –SH functionality for the mercury removal by adsorption process. They achieved 90% of mercury even after seventh cycle among the pool of other metallic ions in wide pH range.[33]

The comparative parameters for both the process are having positive and negative sides. The comparison of the process is to find out the most feasible, greener, and economically viable solution to the contaminant refinement industry. Adsorption techniques are found suitable for the small-scale contaminant refinement operation, but in case of large-scale operations, the most adverse sides of adsorption process are high retention time, extra energy, and chemicals required for the regeneration of the adsorbent materials. While the MOF-based separation membranes are advantageous because of its multicyclic capabilities with high efficiency and selectivity. It also offers the concentrated waste collection facility by backwashing of membranes. At the same time, it has also some limitations in terms of the membrane fouling and cost.

5 Conclusion and Future Perspectives

MOF-based MMPMs are rapid and long-lasting separation material solutions that exhibit the great selectivity, extremely adaptable pore structures, and wide range of application access through the pore tuning of MOF by linker modification or the post-synthetic treatment of MOF materials. MOF-based membranes are highly water-stable materials, and in recent reports, they are widely employed in wastewater treatment and water regeneration. MOF-based materials are found remarkable in separation capabilities. To improve the separation performance of any material, it is necessary to describe the constituents of MOF-based MMPMs. In this chapter, we described the synthetic approaches of MOF materials, bare MOF membranes, and mixed-matrix MOF membranes. In addition to that, this chapter also gives insight to the recent advancement in the field of MOF-based MMPMs and its separation application overview like gas separation, pervaporation, desalination, and most important oily waste water treatment.

In the exploration of MOF-based MMPMs that can be stable and last long term in harsh climatic conditions that are mandatory at present. It is also desired that the optimum solutions are cost-effective too. High cost of the MOF-based MMPMs is the challenge for the commercialisation or large-scale application in water refinement, but there are numerous approaches required for the MOF-based MMPMs' practical applicability.

Acknowledgements DBK and KUP acknowledge PDEU and SODH, respectively, for research fellowship. TKP gratefully acknowledges financial support received from the SERB (TAR/2021/000090).

Notes

1. Johnson N, Revenga C, Echeverria J (2001) Managing water for people and nature. Science 292(5519):1071–1072.
2. Kanzariya DB, Goswami R, Muthukumar D, Pillai RS, Pal TK (2002) Highly luminescent MOF and its in situ fabricated sustainable corn starch gel composite as a fluoro-switchable reversible sensor triggered by antibiotics and oxo-anions. ACS Appl Mater Interfaces 14(43):48658–48674.

3. Shannon MA, Bohn PW, Elimelech M, Georgiadis JG, Mariñas BJ, Mayes AM (2008) Science and technology for water purification in the coming decades. Nature 452(7185):301–310.
4. Koros WJ, Zhang C (2017) Materials for next-generation molecularly selective synthetic membranes. Nat Mater 16(3):289–297.
5. Park HB, Kamcev J, Robeson LM, Elimelech M, Freeman BD (2017) Maximizing the right stuff: the trade-off between membrane permeability and selectivity. Science 356(6343):eaab0530.
6. Li X, Liu Y, Wang J, Gascon J, Li J, Van der Bruggen B (2017) Metal–organic frameworks-based membranes for liquid separation. Chem Soc Rev 46(23):7124–7144.
7. Goh P, Ismail A (2018) A review on inorganic membranes for desalination and wastewater treatment. Desalination 434:60–80.
8. Safaei M, Foroughi MM, Ebrahimpoor N, Jahani S, Omidi A, Khatami M (2019) A review on metal–organic frameworks: synthesis and applications. Trends Anal Chem 118:401–425.
9. Kanzariya DB, Khan TS, Das S, Lama P, Bandyopadhyay R, Pal TK (2022) Highly regenerative, fast colorimetric response for organo-toxin and oxo-anions in an aqueous medium using a discrete luminescent Cd(II) complex in a heterogeneous manner with theoretical revelation. Dalton Trans 51:7436–7454.
10. Li W, Zhang Y, Li Q, Zhang G (2015) Metal–organic framework composite membranes: synthesis and separation applications. Chem Eng Sci 135:232–257.
11. Zhu Y, Gupta KM, Liu Q, Jiang J, Caro J, Huang A (2016) Synthesis and seawater desalination of molecular sieving zeolitic imidazolate framework membranes. Desalination 385:75–82.
12. Kwon HT, Jeong H-K (2013) In situ synthesis of thin zeolitic–imidazolate framework ZIF-8 membranes exhibiting exceptionally high propylene/propane separation. J Am Chem Soc 135:10763–10768.
13. Liu T-Y, Yuan H-G, Liu Y-Y, Ren D, Su Y-C, Wang X (2018) Metal–organic framework nanocomposite thin films with interfacial bindings and self-standing robustness for high water flux and enhanced ion selectivity. ACS Nano 12:9253–9265.
14. Zhu J, Qin L, Uliana A, Hou J, Wang J, Zhang Y (2017) Elevated performance of thin film nanocomposite membranes enabled by modified hydrophilic MOFs for nanofiltration. ACS Appl Mater Interfaces 9:1975–1986.
15. Abbasi AR, Karimi M, Daasbjerg K (2017) Efficient removal of crystal violet and methylene blue from wastewater by ultrasound nanoparticles Cu-MOF in comparison with mechanosynthesis method. Ultrason Sonochem 37:182–191.
16. Serra-Crespo P, Ramos-Fernandez EV, Gascon J, Kapteijn F (2011) Synthesis and characterization of an amino functionalized MIL-101 (Al): separation and catalytic properties. J Membr Sci 23:2565–2572.
17. Li Q, Cheng L, Shen J, Shi J, Chen G, Zhao J (2017) Improved ethanol recovery through mixed-matrix membrane with hydrophobic MAF-6 as filler. Sep Purif Technol 178:105–112.
18. Benzaqui M, Semino R, Carn F, Tavares SR, Menguy N, Giménez-Marqués M (2019) Covalent and selective grafting of polyethylene glycol brushes at the surface of ZIF-8 for the processing of membranes for pervaporation. ACS Sustainable Chem Eng 7:6629–6639.
19. Van Assche TR, Desmet G, Ameloot R, De Vos DE, Terryn H, Denayer JF (2012) Electrochemical synthesis of thin HKUST-1 layers on copper mesh. Microporous Mesoporous Mat 158:209–213.
20. Lee Y-R, Kim J, Ahn W-S (2013) Synthesis of metal–organic frameworks: a mini review. Korean J Chem Eng 30:1667–1680.
21. Ying Y, Liu D, Zhang W, Ma J, Huang H, Yang Q (2017) High-flux graphene oxide membranes intercalated by metal–organic framework with highly selective separation of aqueous organic solution. ACS Appl Mater Interfaces 9:1710–1718.
22. Zhu Y, Gupta KM, Liu Q, Jiang J, Caro J, Huang A (2016) Synthesis and seawater desalination of molecular sieving zeolitic imidazolate framework membranes. Desalination 385:75–82.
23. Qiu S, Xue M, Zhu G (2014) Metal–organic framework membranes: from synthesis to separation application. Chem Soc Rev 43:6116–6140.

24. Deng Y, Wu Y, Chen G, Zheng X, Dai M, Peng C (2021) Metal–organic framework membranes: recent development in the synthesis strategies and their application in oil–water separation. Chem Eng J 405:127004.
25. Herm ZR, Swisher JA, Smit B, Krishna R, Long JR (2011) Metal–organic frameworks as adsorbents for hydrogen purification and precombustion carbon dioxide capture. J Am Chem Soc 133:5664–5667.
26. Lee D-J, Li Q, Kim H, Lee K (2012) Preparation of Ni-MOF-74 membrane for CO_2 separation by layer-by-layer seeding technique. Microporous Mesoporous Mater 163:169–177.
27. Pan Y, Lai Z (2011) Sharp separation of C_2/C_3 hydrocarbon mixtures by zeolitic imidazolate framework-8 (ZIF-8) membranes synthesized in aqueous solutions. Chem Commun 47:10275–10277.
28. Kong J, Li, K (1999) Oil removal from oil-in-water emulsions using PVDF membranes. Sep Purif Technol 16:83–93.
29. Chakrabarty B, Ghoshal AK, Purkait MK (2008) Ultrafiltration of stable oil-in-water emulsion by polysulfone membrane. J Membr Sci 325:427–437.
30. Zhang X, Zhao Y, Mu S, Jiang C, Song M, Fang Q (2018) UiO-66-coated mesh membrane with underwater superoleophobicity for high-efficiency oil–water separation. ACS Appl Mater Interfaces 10:17301–17308.
31. Song M, Zhao Y, Mu S, Jiang C, Li Z, Yang P, Qiu S (2019) A stable ZIF-8-coated mesh membrane with micro-/nano architectures produced by a facile fabrication method for high-efficiency oil–water separation. Sci China Mater 62:536–544.
32. Yang Y, Guo Z, Huang W, Zhang S, Huang J, Yang H, Gu S (2020) Fabrication of multi-functional textiles with durable antibacterial property and efficient oil–water separation via in situ growth of zeolitic imidazolate framework-8 (ZIF-8) on cotton fabric. Appl Surf Sci 503:144079.
33. Li J, Liu Y, Ai Y, Alsaedi A, Hayat T, Wang X (2018) Combined experimental and theoretical investigation on selective removal of mercury ions by metal organic frameworks modified with thiol groups. Chem Eng J 354:790–801.

Control of Dissolved Oxygen in Wastewater Treatment Plant Using NN Adaptive PID Controller

S. M. Tharani, A. Ganesh Ram, and M. Vijayakarthick

Abstract The concentration of dissolved oxygen (DO) within the aeration tank(s) of an activated sludge system is one among the foremost important process control parameters. To maintain the DO level in the aeration tank(s) is one of the controlled parameters in wastewater treatment plant. In this paper, the neural network-based adaptive PID algorithm is proposed to control the DO level to neutralize an activated sludge process-based wastewater treatment. The NNPID control algorithm has good tracking, anti-disturbance, and strong robustness performances as compared to the conventional PID controller.

Keywords Wastewater · NNPID · DO concentration · MATLAB

1 Introduction

In industries, dissolved oxygen (DO) is used to treat the wastewater. Generally, oxygen level is low in wastewater. Therefore, optimum amount of oxygen should be sent to the wastewater. Both excessively low and excessively high levels of dissolved oxygen are often equally as harmful to aquatic life. If DO content goes high above the desired level, energy for supplying the DO is wasted. So the controlling of DO level in the water is necessary. A neural network is being used, which has a basic network topology and good generalization ability in addition to strong self-learning and adaptable characteristics. In several other domains, the proposed network already includes research and application foundations for the control of real operations. The suggested neural network-based PID (NNPID) control algorithm combines the advantages of these two methods when compared to the traditional PID control algorithm. It is straightforward, simple to execute, and provides greater control accuracy. In other words, it eliminates the problem with standard PID controllers that make it impossible to alter parameters online.

S. M. Tharani · A. G. Ram (✉) · M. Vijayakarthick
Department of Instrumentation Engineering, Madras Institute of Technology, Anna University, Chennai 600044, India
e-mail: agram72@gmail.com

N. M. Rao et al. (eds.), *Advanced Nanomaterials and Their Applications*,
Springer Proceedings in Materials 22, https://doi.org/10.1007/978-981-99-1616-0_14

According to a 1998 study published in "Control Engineering," single-loop controllers account for 64% of all applications, whereas multi-loop controllers account for 36%. Classical feedback control systems account for 85% of the total, with feed-forward control accounting for 6% and cascade control accounting for 9%. Despite their simple form, PID controllers are by far the most commonly utilized controllers in control systems. They can be found in both DCS and PLC systems, which are products of evolving technologies and automation and control systems. The majority of the control techniques (cascade control, feed forward control, and so on) were created using the PID control system paradigm. Commercial PID controllers with auto-tuning features have been in use since the early 1980s [1], thanks to advances in microelectronics that allowed for the incorporation of the programs required for auto-tuning. The development of novel estimating and auto-tuning techniques and methodologies [2] is one of the reasons for the rising interest in PID controllers. Another reason is the ability of utilizing predictive control with PID controllers in their main loop based on monitoring state models [3]. It is commonly known that applying predictive control improves the performance of complicated systems significantly. The present document presents in points of interest the ultimate state of Benchmark Simulation Model no. 1 (BSM1). To demonstrate conditions to be executed for the proposed format, the method to test the execution and the execution criteria to be utilized is described, as well as the sensors and control handles. At last open-loop and closed-loop come about gotten with a Matlab-Simulink and FORTRAN execution are proposed [4]. Artificial intelligence approaches for complicated nonlinear systems have recently captured the interest of researchers. Adaptive fuzzy control, which can deal with model uncertainty, was used to regulate DO in [5]. Neural networks are prominent in nonlinear adaptive control because of their capacity to learn huge nonlinearities of the system, and a back-propagation neural network is employed for DO control in [6]. In addition, integrated strategies for DO control have been developed, such as neural-fuzzy control [7] and self-organizing radial basis function neural network model predictive control method [8]. WWTP has widely used model predictive control (MPC) to implement the DO tracking control, and the second simplified model is used [9]. Simulation results show that the model predictive control (MPC) method provides better control performance compared with the PID method. Despite the fact that an analytical process model is needed and the computation load is heavy, a variety of intelligent methods have also been applied to the WWTP, mainly using neural networks (NNs) and fuzzy technology. An adaptive controller for DO concentration based on dynamic structure neural networks [10] is proposed. This paper presents a self-organizing fuzzy control (SOFC) method that incorporates flexible fuzzy rules for the control of DO concentration in real time. The control system consists of a fuzzy controller with a fuzzy neural network (FNN) that learns both the structure and parameters simultaneously, as well as a compensation controller [11]. The concept of artificial neural networks, or ANNs, was originally inspired by the fact that humans are capable of performing multiple complex tasks with relative and is often used to model complex relationships between inputs and outputs. A number of types of ANNs are used in both the academic and industrial sectors, including radial basis function networks (RBFs) [12], feed-forward neural networks (FNNs), and dynamic neural networks [13]. This paper briefly presents the fundamental WWTP and creates a decreased demonstration of DO concentration. At that point, the control objective and RBF NNs are displayed. An versatile NN control plan based on a unsettling influence eyewitness, and by Lyapunov strategy, the confirmation of consistently bounded characteristics of the closed loop framework signals and the compounded unsettling influence gauge error [14].

In this paper, a neural network-based PID control algorithm is proposed. This paper organized into three sections. Sect. 2 consists of implementation of dissolved oxygen control in MATLAB, by using traditional PID method. Sect. 3 consists of design and implementation of PID parameters in neural network. Sect. 4 consists of results and discussion.

2 Implementation of Dissolved Oxygen Control in MATLAB

In aeration processes, improvements in dissolved oxygen sensing technologies are providing an opportunity for accurate measurement as well as significant energy and maintenance cost savings. Most current facilities employ an activated sludge system, which feeds on the organic elements in the sewage with a culture of various microorganisms. These bacteria and organisms use dissolved oxygen to break down organic carbons into carbon dioxide, water, and energy, removing hazardous compounds from the water. The efficacy of the aeration process depends on precise regulation of dissolved oxygen levels. Continuously monitoring dissolved oxygen levels is the most effective technique of ensuring optimum aeration efficiency.

- DO is important for performing BOD removal and nitrification of activated sludge plants. Nitrification growth depends on the DO concentration.
- If DO is high, it decreases the efficiency of de-nitrification process. And also encourage the growth of unwanted filamentous bacteria.
- If DO is low, improper nitrification and it will produce order.

Figure 1 shows the benchmark simulation layout (BSM1), which was supported by the ASM1. It is a rather simple layout like ASM1, the fundamental component of BSM1 is a biological (or biochemical) activated sludge reactor, which is made up of five compartments, two of which are hypoxic tanks, whereas the remaining three are aerobic tanks. The second a part of BSM1 could be a secondary settler. Reactors 1 and 2 are unalternated in open-loop, however completely mixed; reactors 3, 4, and 5 are aerated. For the open-loop case, the oxygen switch coefficients (K_La) are constant for reactors 3 and 4; the coefficients $(K_La)_3$ and $(K_La)_4$ are ready to a constant, in which, the approach of the air waft charge of the blower is constant. For reactor 5, the coefficient $(K_La)_5$ is chosen because the managing variable (or operational variable) on this paper is manipulated for keeping the DO attention at a stage of 3 mg/L (refer Fig. 2).

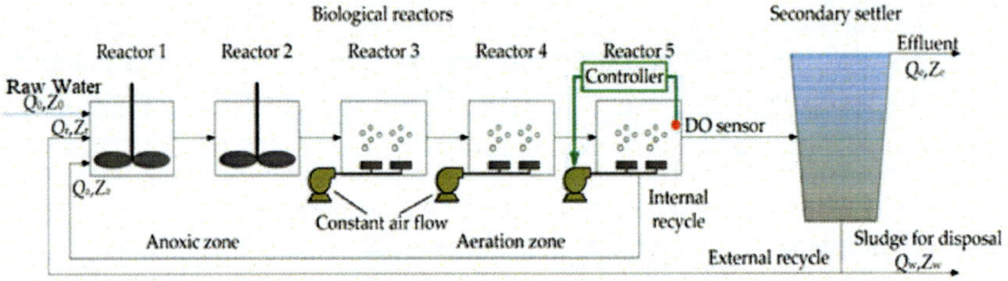

Fig. 1 Benchmark simulation layout (BSM1)

Fig. 2 Dissolved Oxygen Control at Reactor 5

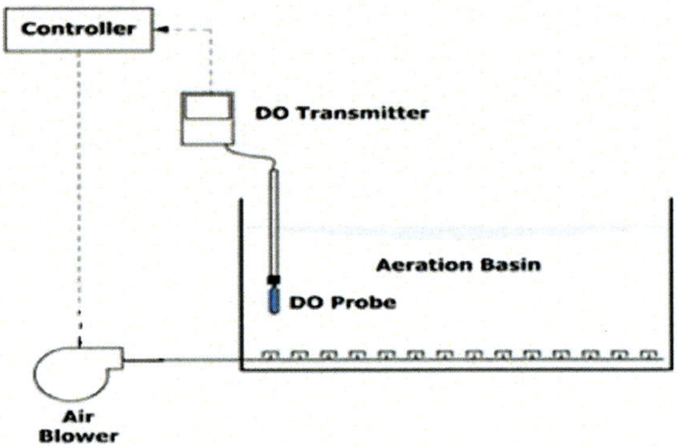

Dissolved oxygen equation is given as

$$\frac{dS_{O,k}}{dt} = \frac{1}{V_k}\left(Q_{k-1}S_{O,k-1} + r_k V_k (K_L a)_k V_k \left(S_O^* - S_{O,k}\right) - Q_k S_{O,k}\right)$$

where Q is the flow rate, V is the volume of the reactor, r is the reaction rate, and k is the reactor

$K_L a$ is the oxygen transfer coefficient, S_O is the dissolved oxygen concentration, and S_O^* is the saturation concentration for oxygen ($S_O^* = 8$ g/m^3 at 15 °C).

The parameter values are shown in Table 1.

Dissolved oxygen equation for reactors:

For Reactor 1,

$$\frac{dS_{O,1}}{dt} = \frac{1}{V_1}\left(Q_0 S_{O,0} + r_1 V_1 (K_L a)_1 V_1 \left(S_O^* - S_{O,1}\right) - Q_1 S_{O,1}\right) \quad (1)$$

For Reactor 2,

$$\frac{dS_{O,2}}{dt} = \frac{1}{V_2}\left(Q_1 S_{O,1} + r_2 V_2 (K_L a)_2 V_2 \left(S_O^* - S_{O,2}\right) - Q_2 S_{O,2}\right) \quad (2)$$

Table 1 Parameter values

Symbol	Parameters	Values	Unit
Q_{k-1}	Previous reactor flow rate	92,230	m^3/day
$S_{O,k-1}$	Previous reactor S_O level	1.72	mg/L
Q_k	Present reactor flow rate	92,230	m^3/day
$S_{O,k}$	Present reactor S_O level	2.43	mg/L
r_k	Reaction rate	0.8	h^{-1}
V_K	Volume of the reactor	1000	m^3
S_O^*	Saturation S_O level	8	g/m^3
$(K_La)_k$	Oxygen transfer coefficient	10	h^{-1}

For Reactor 3,

$$\frac{dS_{O,3}}{dt} = \frac{1}{V_3}\left(Q_2 S_{O,2} + r_3 V_3 (K_L a)_3 V_3 \left(S_O^* - S_{O,3}\right) - Q_3 S_{O,3}\right) \tag{3}$$

For Reactor 4,

$$\frac{dS_{O,4}}{dt} = \frac{1}{V_4}\left(Q_3 S_{O,3} + r_4 V_4 (K_L a)_4 V_4 \left(S_O^* - S_{O,4}\right) - Q_4 S_{O,4}\right) \tag{4}$$

For Reactor 5,

$$\frac{dS_{O,5}}{dt} = \frac{1}{V_5}\left(Q_4 S_{O,4} + r_5 V_5 (K_L a)_5 V_5 \left(S_O^* - S_{O,5}\right) - Q_5 S_{O,5}\right) \tag{5}$$

2.1 Open-Loop Response

The open-loop response for Rectors 1–5 is shown in Fig. 3.

2.2 PID Controller Design

The PID controller is based on by using Ziegler–Nicholas tuning method. The output of concentration of dissolved oxygen is maintained at 3 mg/L. The PID controller takes the corrective action and maintains the S_O level at 3 mg/L. The closed-loop response is shown in Fig. 4.

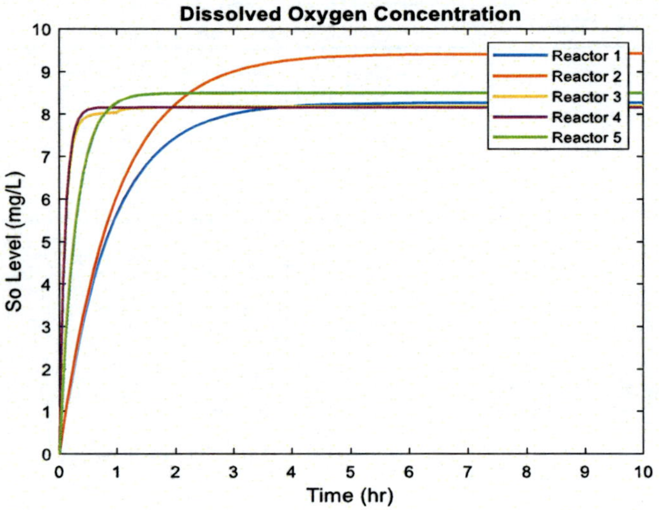

Fig. 3 Open-loop responses for rectors 1–5

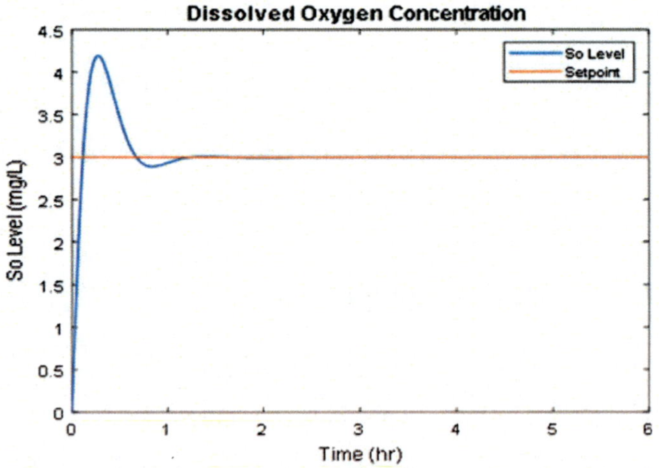

Fig. 4 DO Level control using PID controller

2.3 Implementation of PID Parameters Using Neural Network

An artificial neural network (refer Fig. 5) is a machine learning system that simulates the behavior of biological neural networks (BNNs). It uses a nonlinear processing unit to model biological neurons and adjusts the variable weights between connected

Fig. 5 General architecture of artificial neural network

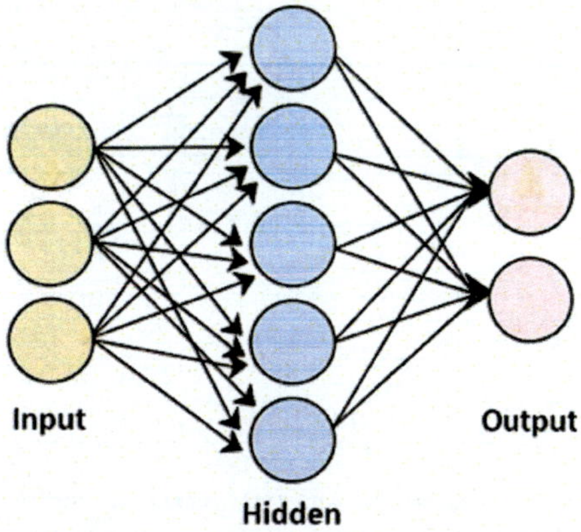

units to simulate the behavior of biological synapses among neurons. The network's distinctive topological structure is organized in a certain connected manner from each processing unit. The main features of ANN are parallel processing and distributed storage. It also has high fault tolerance and nonlinear mapping capabilities, as well as self-organization, self-learning, and adaptive reasoning.

In most cases, especially in deep learning, the neurons are grouped into many layers. Neurons in one layer are only connected to neurons in the layers immediately before and following it. The layer that gets outside information is the info layer. The layer that delivers the final word results is the result layer. In the middle of them are at least zero secret layers. Single-layer and unlayered networks are utilized. Between two layers, various association designs are conceivable. They will be completely associated, with each neuron in one layer interfacing with every neuron inside the following layer. They will pool, where a lot of neurons in a single layer associate with one neuron inside the following layer, accordingly decreasing how much neurons in that layer. Neurons with just such associations structure a coordinated non-cyclic chart and are alluded to as take care of forward networks. Then again, networks that permit associations between neurons inside the equivalent or past layers are alluded to as intermittent organizations.

Steps Involved in designing Neural Network-based PID Controller:

The neural network is designed in MATLAB 2021a, and the command is "nnstart".

- Assemble the training data:

The input data from reactor 5 and PID controller command signal are uploaded into the input data. The neural network structure is shown in figure. Here the number of input is 2 and output is 3. Based on the error values, we can adjust the hidden layer.

- Create the network object:

A two-layer feed-forward network with sigmoid secret neurons and direct result neurons can fit complex planning issues self-assertively well.

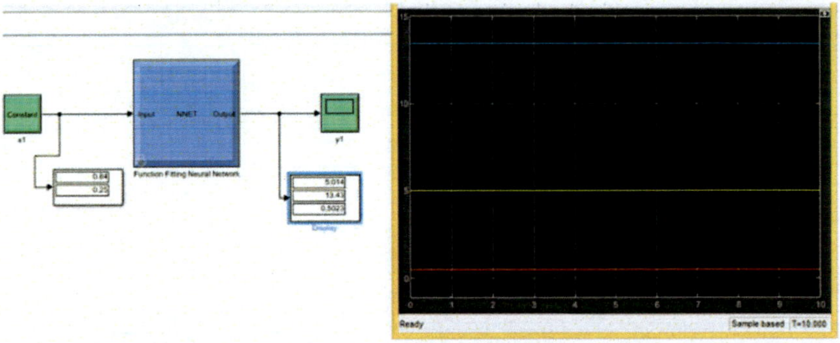

Fig. 6 PID parameter generation using Neural Network tool in MATLAB

- Train the network:

Levenberg–Marquardt training algorithm is used. This algorithm typically requires more memory but less time. In batch training, weights and biases are only updated after all the inputs and targets are presented. For layer 1, tansig transfer function is used. Multilayer networks may use this transfer function.

For layer 2, purelin transfer function is used. Linear transfer function is used in back-propagation network.

- Simulate the network response to new inputs:

After the training, validation, and testing, the new controller parameters are K_p, K_i, and K_d are generated. Based on the new controller parameters, the PID controller is designed which is shown in Fig. 6.

The controller parameter values K_p, K_i, and K_d taken for different DO concentration and controller command of that responses are shown in Fig. 7.

3 Result and Discussion

Figure 8 shows the comparison of traditional PID and PID parameters which are generated by neural network. Compared with the conventional PID controller, the NNPID controller can quickly and accurately track the desired output trajectory values, which means it not only has a good tracking performance, but also has a stronger anti-disturbance ability with the changes to the set points. PID may also fail to obtain the manage intention or impact of the technique at the same time as the use of the conventional PID manage set of rules because of unknown and sudden disturbances in addition to substantial modifications in running conditions. The DO concentration is difficult to take care of at point under the control of the traditional incremental PID controller when the influent flow and quality changed greatly. NNPID can effectively maintain the DO concentration round the set value

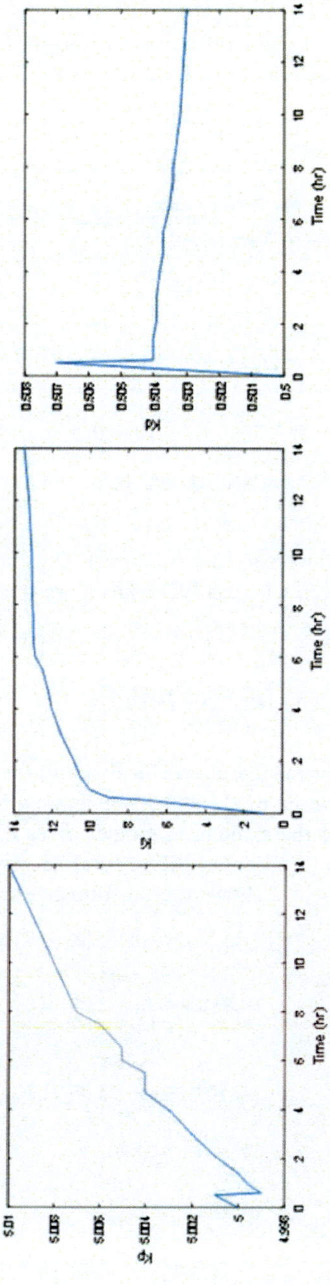

Fig. 7 Controller parameter values K_p, K_i, and K_d

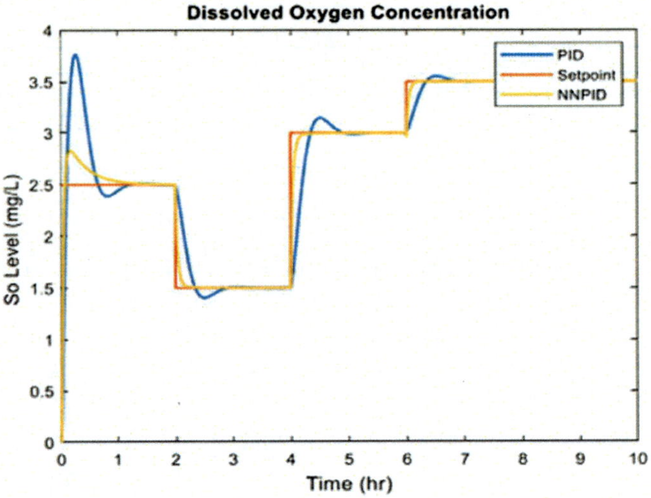

Fig. 8 Servo response of PID and NNPID controllers

with a comparatively low error. Rise time for PID is 0.089 s, for NNPID is 0.065 s, also NNPID settles faster, whereas PID takes more time.

3.1 Performance Evaluation Index

Two main indices for evaluating the performance of the dissolved oxygen controller are Integral of Absolute Error (IAE) and Integral of Squared Error (ISE). Generally, a smaller value of the evaluation indices gives the better performance of the controller. Compared to traditional PID, neural network-based adaptive PID gives better performance. Table 2 shows the performance evaluation index of dissolved oxygen concentration.

$$\mathrm{IAE} = \int_{t=0}^{t=10} |e_i| \mathrm{d}t$$

$$\mathrm{ISE} = \int_{t=0}^{t=10} e_i^2 \mathrm{d}t$$

Table 2 Performance evaluation index

Method	IAE	ISE
PID	0.308	0.0348
NNPID	0.201	0.015

4 Conclusion

In this paper, an adaptive PID control algorithm based on neural network is proposed. The NNPID algorithm combines the good learning and adaptive ability of neural networks and the practical advantages of PID algorithm. The gradient descent method is employed to adaptively adjust the increment of the three parameters of the PID controller to attain an optimal control effect on the control of DO concentration. The simulation results show that the NNPID algorithm not only includes a better performance of tracking and anti-jamming, but also contains a great improvement to the robustness compared to it of the traditional PID. The traits of sturdy coupling, nonlinearity, and large time postponement of dissolved oxygen management device in activated sludge wastewater remedy and managing set of rules referred to as NNPID set of rules are discussed.

References

1. Astrom KJ, Hagglund T (2004) Revisiting the Ziegler–Nichols step response method for PID control. J Process Control 14:635–650
2. Johnson MA, Moradi MH (2005) PID control. Springer-Verlag London Limited
3. Nascu I, et al (2008) Development and application of a PID auto-tuning method to a wastewater treatment process. In: 2008 IEEE international conference on automation. Quality and testing, robotics, vol 2. IEEE
4. Alex J, et al (2008) Benchmark simulation model no. 1 (BSM1). In: Report by the IWA Task group on benchmarking of control strategies for WWTPs, pp 19–20
5. Belchior CAC, Araújo RM, Landeck JAC (2012) Dissolved oxygen control of the activated sludge wastewater treatment process using stable adaptive fuzzy control. Comput Chem Eng 37:152–162
6. Syu M-J, Chen B-C (1998) Back-propagation neural network adaptive control of a continuous wastewater treatment process. Ind Eng Chem Res 37(9):3625–3630
7. Mingzhi H, et al (2009) Control rules of aeration in a submerged biofilm wastewater treatment process using fuzzy neural networks. Expert Syst Appl 36(7):10428–10437
8. Han H-G, Qiao J-F, Chen Q-L (2012) Model predictive control of dissolved oxygen concentration based on a self-organizing RBF neural network. Control Eng Pract 20(4):465–476
9. Holenda B, et al (2008) Dissolved oxygen control of the activated sludge wastewater treatment process using model predictive control. Comput Chem Eng 32(6):1270–1278
10. Han H-G, Qiao J-F (2011) Adaptive dissolved oxygen control based on dynamic structure neural network. Appl Soft Comput 11(4):3812–3820
11. Qiao J, Zhang W, Han H (2016) Self-organizing fuzzy control for dissolved oxygen concentration using fuzzy neural network. J Intell Fuzzy Syst 30(6):3411–3422
12. Wu X-J, et al (2008) Predictive control of SOFC based on a GA-RBF neural network model. J Power Sources 179(1):232–239

13. Zhang X, Hoo KA (2008) Controlling a coupled set of biological reactors for wastewater treatment. Control Eng Pract 16(5):553–568
14. Lin M-J, Luo F (2016) Adaptive neural control of the dissolved oxygen concentration in WWTPs based on disturbance observer. Neuro Comput 185:133–141